ELECTRICITY 3

ELECTRICITY 3

MOTORS & GENERATORS, CONTROLS, TRANSFORMERS

FOURTH EDITION

WALTER N. ALERICH

DELMAR PUBLISHERS INC.®

NOTICE TO THE READER

Publisher does not warrant or guarantee any of the products described herein or perform any independent analysis in connection with any of the product information contained herein. Publisher does not assume, and expressly disclaims, any obligation to obtain and include information other than that provided to it by the manufacturer.

The reader is expressly warned to consider and adopt all safety precautions that might be indicated by the activities described herein and to avoid all potential hazards. By following the instructions contained herein, the reader willingly assumes all risks in connection with such instructions.

The publisher makes no representations or warranties of any kind, including but not limited to, the warranties of fitness for particular purpose or merchantability, nor are any such representations implied with respect to the material set forth herein, and the publisher takes no responsibility with respect to such material. The publisher shall not be liable for any special, consequential or exemplary damages resulting, in whole or in part, from the readers' use of, or reliance upon, this material.

Delmar Staff:

    Administrative Editor: Mark Huth
    Developmental Editor: Marjorie A. Bruce
    Production Editor: Carol Micheli

For information address Delmar Publishers Inc.
2 Computer Drive West, Box 15–015
Albany, New York 12212–5015

Printed in the United States of America
Published simultaneously in Canada
by Nelson Canada,
A division of International Thomson Limited

10 9 8 7 6 5 4

Library of Congress Cataloging in Publication Data

Alerich, Walter N.
   Electricity 3.

   Includes index.
   1. Electric machinery–Direct current. 2. Electric controllers. 3. Electric transformers. I. Title.
II. Title: Electricity three.
TK2612.A43   1986   621.31'042   85–23794
ISBN 0–8273–2533–9 (soft)
ISBN 0–8273–2535–5 (Instructor's guide)

CONTENTS

PREFACE

The fourth edition of ELECTRICITY 3 has been updated to reflect current devices, equipment, and techniques used in the installation of rotating machinery. In addition, the section of the text devoted to transformer theory and applications has been significantly updated. At the same time, the text has retained the features that have made it so popular through previous editions.

The text introduces the basic principles of automatic motor control. Each type of control presented is thoroughly described and well-illustrated. A detailed explanation of the operation is provided and numerous typical schematic and wiring diagrams are given to familiarize students with common installations. This type of thorough explanation better prepares students to perform effectively on the job in the installation, troubleshooting, repair, and service of rotating machinery and its associated controls.

The knowledge obtained by a study of this text permits the student to progress to further study. It should be realized that both the development of the subject of electricity and the study of the subject are continuing processes. The electrical industry constantly introduces new and improved devices and materials, which in turn often lead to changes in installation techniques. Electrical codes undergo periodic revisions to upgrade safety and quality in electrical installations.

The text is easy to read and the topics are presented in a logical sequence. However, the instructor can elect to follow a different sequence depending upon available time and the requirements of individual programs.

Each unit begins with objectives to alert students to the learning that is expected as a result of studying the unit. An Achievement Review at the end of each unit tests student understanding to determine if the objectives have been met. Note that the problems presented in this text require the use of simple algebra only for their solution. Following selected groups of units, a summary review unit contains additional questions and problems that test student comprehension of a block of information. This combination of reviews is essential to the learning process required by this text.

All students of electricity will find this text useful, especially those in electrical apprenticeship programs, trade and technical schools, and various occupational programs.

It is recommended that the most recent edition of the National Electrical Code (published by the National Fire Protection Association) be available for reference as the student uses ELECTRICITY 3. Applicable state and local regulations should also be consulted when making actual installations.

Features of the fourth edition include:

- Updated photos to reflect modern equipment and devices
- Modifications of selected circuit diagrams to include solid-state devices
- Content updated to the requirements of the 1984 National Electrical Code (the most current)
- Significantly revised units in the transformer section with new, practical applications given
- New information on dc variable speed motor drives
- New glossary

A combined Instructor's Guide for ELECTRICITY 1 through ELECTRICITY 4 is available. The guide includes the answers to the Achievement Reviews and Summary Reviews for each text and additional test questions covering the content of the four texts. Instructors may use these questions to devise additional tests to evaluate student learning. Student Study Guides to accompany each text will give students additional opportunities for classroom and laboratory practice.

ABOUT THE AUTHOR

Walter N. Alerich, BVE, MA, has an extensive background in electrical installation and education. As a journeyman wireman, he has many years of experience in the practical applications of electrical work. Mr. Alerich has also served as an instructor, supervisor and administrator of training programs, and is well-aware of the need for effective instruction in this field. A former department head of the Electrical-Mechanical Department of Los Angeles Trade-Technical College, Mr. Alerich has written extensively on the subject of electricity and motor controls. He presently serves as an international specialist/consultant in the field of electrical trades, developing curricula and designing training facilities. Mr. Alerich is also the author of ELECTRICITY 4 and ELECTRIC MOTOR CONTROL, and the coauthor of INDUSTRIAL MOTOR CONTROL.

ACKNOWLEDGMENTS

The revision of ELECTRICITY 3 was based on information and recommendations submitted by the following instructors:

Eugene Gentile and James Richards, Auburn Career Center, Painesville, OH 44077
Mark D. Kocher, Columbus Technical Institute, Columbus, OH 43162
James L. Gettings, Lyons Technical Institute, Philadelphia, PA 19134
Robert F. Smeal, Greater Johnstown Area Vocational Technical School, Johnstown, PA 15905
Richard M. Berube, Licking County Joint Vocational School, Newark, OH 43055
Michael F. Auth, Lyons Institute, Clark, NJ 07066
Sam A. Portaro, Davidson County Community College, Lexington, NC 27292
DeWitt Booth, Southeastern Community College, West Burlington, IA 52655
Charles W. Thompson, J.F. Drake State Technical College, Huntsville, AL 35811
Buck Deaver, Martin Community College, Williamston, NC 27892

The revised manuscript was thoroughly reviewed by the following instructors:

John Close, Dean Institute of Technology, Pittsburgh, PA 15226
Melvin R. Houck, Greenville Technical College, Greenville, SC 29606
Gene A. Hilst, Blackhawk Technical Institute, Janesville, WI 53547
James Brozek, Brazosport College, Lake Jackson, TX 77566

ELECTRICAL TRADES

The Delmar series of instructional material for the basic electrical trades consists of the texts, text-workbooks, laboratory manuals, and related information workbooks listed below. Each text features basic theory with practical applications and student involvement in hands-on activities.

ELECTRICITY 1
ELECTRICITY 2
ELECTRICITY 3
ELECTRICITY 4
ELECTRIC MOTOR CONTROL
ELECTRIC MOTOR CONTROL LABORATORY MANUAL
INDUSTRIAL MOTOR CONTROL
ALTERNATING CURRENT FUNDAMENTALS
DIRECT CURRENT FUNDAMENTALS
ELECTRICAL WIRING – RESIDENTIAL
ELECTRICAL WIRING – COMMERCIAL
ELECTRICAL WIRING – INDUSTRIAL
PRACTICAL PROBLEMS IN MATHEMATICS FOR ELECTRICIANS

EQUATIONS BASED ON OHM'S LAW

P = Power in Watts
I = Intensity of Current in Amperes
R = Resistance in Ohms
E = Electromotive Force in Volts

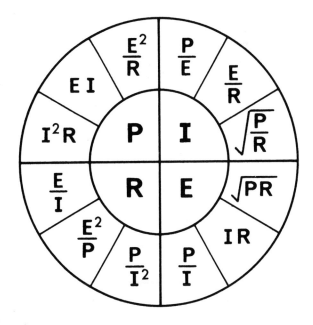

UNIT OPERATING PRINCIPLES OF DC GENERATORS

OBJECTIVES

After studying this unit, the student will be able to

- state the function of a dc generator.
- list the major components of a generator.
- describe the difference between a separately-excited and a self-excited generator.
- explain how the output voltage of a generator can be varied.

A *dc generator* changes mechanical energy into electrical energy. It furnishes electrical energy only when driven at a definite speed by some form of prime mover.

Dc generators are used principally in electrical systems for mobile equipment. They are also used in power plants supplying dc power for factories and in certain railway

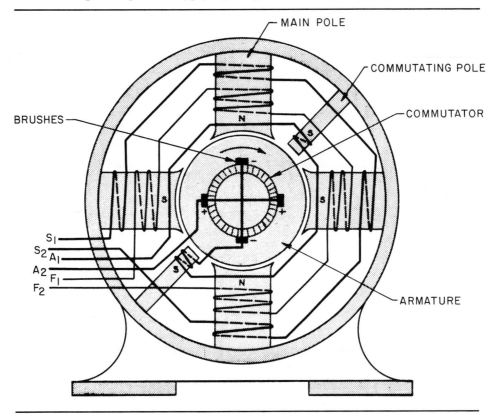

Fig. 1-1 Compound generator fields, with commutating poles

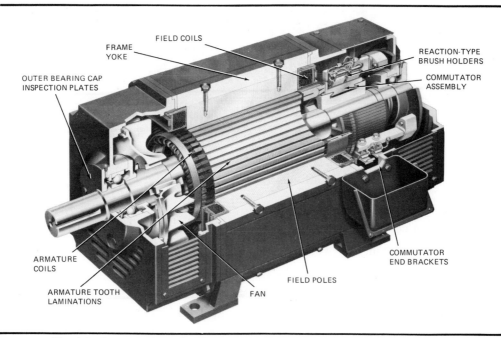

Fig. 1-2 Cutaway view of a direct-current generator *(Courtesy Reliance Electric)*

systems. Dc power is used extensively in communication systems and for battery charging and electroplating operations. The generation of electromotive force is described in detail in ELECTRICITY 1.

DC GENERATOR COMPONENTS

The essential parts of a dc generator are shown in figures 1-1 and 1-2. The member which is rotated is called the armature. An *armature* is a cylindrical, laminated iron structure mounted on a drive shaft. An *armature winding* is embedded in slots on the surface of the armature. Voltage is induced in this winding. The winding itself consists of a series of loops. The ends of these loops terminate at the copper segments of a commutator.

A *commutator* consists of a series of copper segments which are insulated from one another and the shaft. The commutator turns with the shaft and the armature windings. The commutator is used to change the ac voltage induced in the armature windings to dc voltage at the generator output terminals. Carbon brushes pressing against the commutator segments lead the current to the external load circuit.

The armature windings generate voltage by cutting a magnetic field as the armature rotates. This magnetic field is established by electromagnets mounted around the periphery of the generator. The electromagnets, called *field poles,* are arranged in a definite sequence of magnetic polarity; that is, each pole has a magnetic polarity opposite to that of the field poles adjacent to it. Electrical power for the generator field circuit is usually obtained from the generator itself.

When a generator feeds a load circuit, current passing through the armature sets up a magnetic field around the armature. This field reacts with the main field flux. The result is a force acting to turn the armature in a direction opposite to that in which it is being driven (figure 1-3). The force of this reaction is proportional to the current in the armature

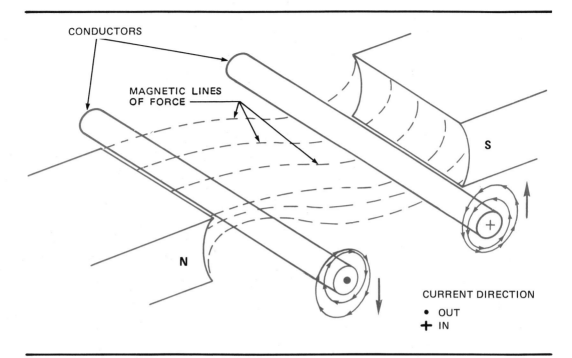

Fig. 1-3 "Motor action" opposing generator driving force

and accounts for the fact that more power is needed to drive a generator when electrical energy is taken from it.

Armature Reaction

The armature field flux also reacts against the main field flux and tends to distort it. One result of this undesirable condition, known as *armature reaction*, is excessive sparking at the brushes on the commutator. To counteract this effect, commutating poles are often inserted between the main field poles, as shown in figure 1-1. These commutating poles, also called *interpoles*, are energized by windings placed in series with the output (load) circuit of the generator. Because of this arrangement, armature reaction, which tends to increase with load current, is counteracted by the effects of this same increase of current.

Armature reaction, appearing as excessive brush sparking under load, also can be partially corrected by shifting the brushes from their neutral position in the direction of rotation. Large dc generators have the brushes assembled so that they can be shifted to the position of minimum sparking. When the brushes are not movable, the generator manufacturer inserts other design features to minimize the effects of armature reaction.

Brush Polarity

The output terminals of a generator, as with other dc power units, have electrical polarity. In the case of generators, the term *brush polarity* is used to distinguish between the *electrical polarity* of the brushes and the *magnetic polarity* of the field poles.

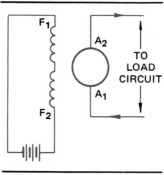

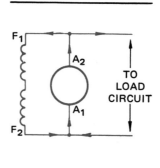

Fig. 1-4 Separate excitation   Fig. 1-5 Self-excitation

Brush polarity markings are often omitted, but the electrician can easily determine electrical polarity by connecting a voltmeter across the output leads of the generator. Many automotive and aircraft generators are constructed with either the positive or negative brushes grounded to the frame of the generator. It is very important to maintain the polarity as specified by the manufacturer. Additional information on brush polarity will be given after the effects of residual magnetism in the field circuit are considered.

Field Supply

The magnetic field of a generator is established by a set of electromagnets (field poles). The power required by the field circuit may be supplied from a separate dc supply. If this is the case, the generator is said to have a *separately excited field*. The majority of generators, however, are self-excited and the power for the field is supplied by the generator itself.

Figure 1-4 illustrates a separately excited dc generator with the field circuit supplied from batteries. A self-excited shunt generator is shown in figure 1-5. Note that the field circuit is connected in parallel with the armature and that a small part of the generator output is diverted to the field circuit.

OUTPUT VOLTAGE CONTROL

Since the induced voltage depends on the rate at which the magnetic flux is cut, it is possible to vary the output voltage by controlling the speed of the prime mover or the strength of the magnetic field. In all but a few instances, the output voltage is controlled by varying the field current with a rheostat in the field circuit.

The flux density in the field poles depends on the field current. As a result, the voltage output of the generator continues to increase with an increase of field current to a point where saturation of the field poles occurs. Any additional increase in voltage output after this point must be obtained by an increase in speed.

GENERATOR RATINGS

Generator ratings as specified by the manufacturer are usually found on the nameplate of the machine. The manufacturer generally specifies the kilowatt output, current, terminal voltage, and speed of the generator. For large generators, the ambient temperature is also given.

ROTATION

A separately-excited generator develops voltage for either direction of rotation. This is not true, however, for self-excited units; they develop voltage in one direction only. The standard direction of rotation for dc generators is clockwise when looking at the end of the generator opposite the drive shaft (this is usually the commutator end).

REGULATION

The voltage regulation of a generator is one of its important characteristics. Different types of generators have different voltage regulation characteristics.

Figure 1-6 shows the action of the voltage at the terminals of a generator for different values of the load current. The drop in terminal voltage is caused by the loss in voltage (1) across the internal resistance of the armature circuit including the brush contacts, and (2) due to armature reaction. The curve at (a) is the normal curve for a shunt generator. An ideal condition is shown in (b) where the voltage remains constant with load current. Curve (c) illustrates a generator with very poor regulation in that the output voltage drops off considerably as the load current increases. A rising characteristic, curve (d), is obtained by using a cumulative compound-wound generator (unit 4).

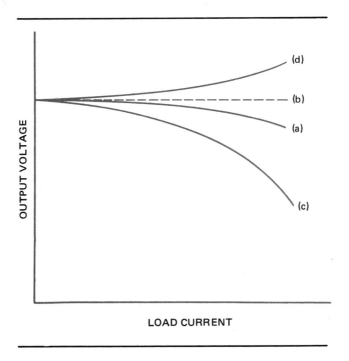

Fig. 1-6 Voltage regulation graphs

ACHIEVEMENT REVIEW

Select the correct answer for each of the following statements and place the corresponding letter in the space provided.

1. A generator   _____
    a.   changes electrical energy to mechanical energy.
    b.   changes mechanical energy to electrical energy.
    c.   is always self-excited.
    d.   is always separately-excited.

2. One of the following is not essential in generating a dc voltage: _____
    a.   a magnetic field   c.   sliprings
    b.   a conductor   d.   relative motion

3. Commutating poles are   _____
    a.   fastened to the center of the commutator.
    b.   located midway between the main poles.
    c.   secondary poles induced by cross magnetizing the armature.
    d.   used to regulate the voltage at the armature.

4. The winding on an interpole is   _____
    a.   made of many turns of fine wire.
    b.   wound in a direction opposite to that of the armature winding.
    c.   connected in series with the armature load.
    d.   connected across the generator terminals.

5. Generator terminals A_1 and A_2 are terminals which   _____
    a.   connect the armature only.
    b.   connect the shunt field in series with the armature.
    c.   connect the series field to the armature.
    d.   have the armature in parallel with the commutating poles.

6. To raise generator voltage, the   _____
    a.   field current should be increased.
    b.   field current should be decreased.
    c.   speed should be decreased.
    d.   brushes should be shifted forward.

7. Generator voltage output control is usually accomplished by   _____
    a.   varying the speed.
    b.   a rheostat in the field.
    c.   increasing the flux.
    d.   decreasing the flux.

8. In figure 1-6, the normal voltage regulation for a shunt generator
    is at   _____
    a.   curve a.   c.   curve c.
    b.   a broken line.   d.   curve d.

THE SEPARATELY-EXCITED DC GENERATOR

OBJECTIVES

After studying this unit, the student will be able to

- explain the relationship of field current, field flux, and output voltage for a separately-excited dc generator.
- describe the effects on the brush polarity of reversing the armature rotation and the field current.
- define residual flux and residual voltage.
- draw and explain the basic circuit.
- connect the generator.

The separately-excited dc generator has few commercial applications, but a knowledge of its operations is an excellent background for understanding other types of generators.

Figure 2-1 represents a separately-excited dc generator supplying power to a load circuit. A current, called the *field current,* passes through the field circuit as soon as switch S_1 is closed. The field current is controlled by the field rheostat, a variable resistor in this case. The amount of *field flux* depends on the field current; however, the field flux is limited by the fact that the iron cores of the field poles become saturated at a definite value of field current. The direction of the field flux is controlled by the direction of the field current.

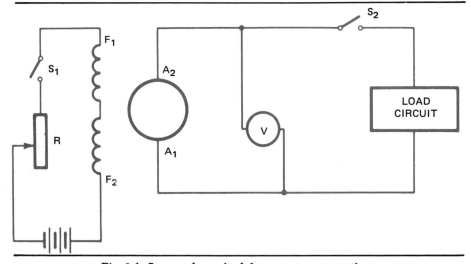

Fig. 2-1 Separately-excited dc generator connections

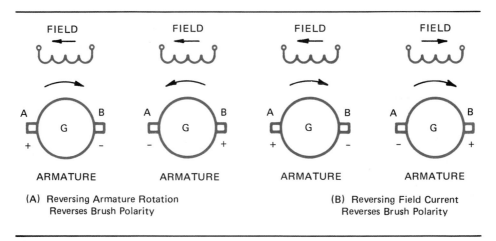

(A) Reversing Armature Rotation
     Reverses Brush Polarity

(B) Reversing Field Current
     Reverses Brush Polarity

Fig. 2-2 Factors affecting brush polarity

The output voltage of the generator is developed as an induced voltage in the armature conductors. This induced voltage appears across the brushes and the generator output terminals designated as A_1 and A_2 in figure 2-1. The output voltage varies with the speed of rotation of the armature and the strength of the field flux.

BRUSH POLARITY

When the armature is driven in either direction, an electrical polarity is established at the generator output terminals and at the brushes. If the machine is stopped and then driven in the opposite direction, the field flux is cut in the opposite direction and the brush polarity changes, as in figure 2-2A.

If the direction of rotation is not changed and the field current is reversed, the same effect is obtained; that is, if the armature conductors maintain a rotation in one direction and cut flux established in the opposite direction, then the brush polarity also changes, as in figure 2-2B.

As a result, the brush polarity in a separately-excited generator can be changed by reversing the rotation of the armature or the direction of the field current. However, if both the armature direction and field current change, the brush polarity would remain the same (unchanged).

OUTPUT VOLTAGE

The magnitude of the voltage depends on the rate at which the flux is cut. In a separately-excited generator, an output voltage increase is proportional to an increase in the armature speed. The upper limit of the voltage is determined by the permissible speed and the insulation qualities of the armature and the commutator.

The output voltage of a separately-excited generator can be varied by adjusting the speed of the armature rotation or the field current. A change in speed always results in a corresponding change in output voltage. An increase in field current increases the output voltage only if the field poles are not saturated. Field control of the output voltage is accomplished by varying the total resistance of the field circuit with a field rheostat, as shown in figure 2-1.

RESIDUAL VOLTAGE

If the field circuit is opened at S_1 (figure 2-1) the field current becomes zero. A small amount of magnetic flux called *residual flux* remains. The small voltage generated when the armature cuts this flux is called *residual voltage*. Brush polarity remains the same when the field current is zero because the residual flux has the same direction as the main flux. If the armature is rotated in the opposite direction, the same residual voltage is obtained at the same speed but the brush polarity reverses. If the field circuit is closed momentarily and the battery connections are reversed, the residual flux reverses and the brush polarity reverses.

ACHIEVEMENT REVIEW

A. Select the correct answer for each of the following statements and place the corresponding letter in the space provided.

1. A separately-excited dc generator has the field connected _____
   a. across the armature.
   b. in series with the armature.
   c. to an external circuit.
   d. none of these.

2. F_1 and F_2 generator terminals are _____
   a. shunt field leads.   c. armature leads.
   b. series field leads.   d. commutating pole leads.

3. The voltage of a separately-excited dc generator may be increased by _____
   a. increasing the speed of rotation of the armature.
   b. decreasing the magnetic flux.
   c. both a and b.
   d. neither a nor b.

4. The function of brushes on a generator is to _____
   a. carry the current to the external circuit.
   b. prevent sparking.
   c. keep the commutator clean.
   d. reverse the connections to the armature to provide dc.

5. Electrical polarity at the brushes may be changed by _____
   a. reversing the rotation of the armature.
   b. reversing the direction of the field current.
   c. either a or b.
   d. neither a nor b.

B.   Select the correct answer to questions 6 to 9 from the following list and write it in the space provided.

power source   only one   decrease
always   strength of the field flux   sometimes
speed of the armature   either   increase

6. In addition to armature rotation, the output voltage varies with the _____ .

7. One factor limiting an increase in output voltage is the _____ .

8. A change in the speed of rotation of the armature _____ results in a change in the output voltage.

9. If the field poles are saturated, an increase in the field current does not cause a(an) _____ in the output voltage.

THE SELF-EXCITED SHUNT GENERATOR

OBJECTIVES

After studying this unit, the student will be able to

- identify a self-excited shunt generator from a circuit diagram.
- describe the way in which the voltage buildup occurs for this type of generator.
- list the causes for a failure of the voltage to build up.
- describe three methods which can be used to renew residual magnetism.
- define voltage control and voltage regulation.
- draw the basic circuit.
- connect the generator.

Most dc generators are of the shunt type with self-excitation. A generator is called a *shunt generator* when its field circuit is connected in parallel with the armature and load. In the field circuit, itself, a four-pole winding may be connected in series, parallel, or series-parallel. The circuit arrangement of the field windings does not affect the classification of the generator because the field windings, as a group, are connected in parallel with the armature and load.

VOLTAGE BUILDUP

Figure 3-1 shows the schematic diagram of a self-excited shunt generator. Voltage control is obtained with a field rheostat. Unlike the separately excited generator, there is no current in the field circuit when the armature is motionless. Since a small amount of residual magnetism is present in the field poles, a weak residual voltage is induced in

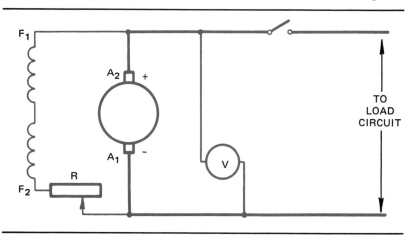

Fig. 3-1 Self-excited shunt generator

the armature as soon as the armature is rotated. This residual voltage produces a weak current in the field circuit. If this current is in the proper direction, an increase in magnetic strength occurs with a corresponding increase in voltage output. The increased voltage output, in turn, increases the field current and the field flux which, again, increase the voltage output. As a result of this action, the output voltage builds up until the increasing field current saturates the field poles. Once the poles are saturated, the voltage remains at a constant level, unless the speed of the armature rotation is changed.

If the direction of armature rotation is reversed, the brush polarity also is reversed. The residual voltage now produces a field current which weakens the residual magnetism and the generator voltage fails to build up. Therefore, a self-excited machine develops its operating voltage for one direction of armature rotation only. The generator load switch may be closed when the desired voltage is reached.

LOSS AND RENEWAL OF RESIDUAL MAGNETISM

A shunt generator may not develop its rated operating voltage due to a loss of residual magnetism. The residual flux may be renewed by momentarily connecting a low-voltage dc source across the field circuit. Several methods can be used to renew the residual magnetism.

Method 1

    a. Disconnect the field circuit leads from the brushes.

    b. Momentarily connect a storage battery or low-voltage dc source to the field circuit leads. To maintain the desired brush polarity, connect the positive terminal of the battery to the field lead normally attached to the positive generator brush.

Method 2

    a. If it is inconvenient to detach the field leads and the brush assembly can be reached, lift either the positive or the negative brush and insert a piece of heavy, dry paper between the brush and the commutator segments.

    b. Momentarily connect a battery to the output leads. With the brush lifted, current passes through the field circuit only. (To maintain the original brush polarity, connect the positive terminal of the battery to the positive generator output terminal.)

    c. Remove the paper under the brush before restarting the generator.

Method 3

    a. If it can be done readily, disconnect the generator from its prime mover.

    b. Then, restore the residual field by momentarily connecting a battery to the generator output leads. Since the field circuit is connected across the output leads, the current renews the magnetic field.

Caution: If the armature is not free to rotate, damage to the armature assembly may occur.

When the battery voltage is high enough in Method 3, the generator armature rotates as a motor. The rotation produced does not contribute to restoring the residual flux. However, this effect, called *motorizing*, is useful because it is a rough check of the overall generator operation. That is, the armature should rotate freely if the voltage applied is a sizable fraction of the rated output voltage, with the direction of armature rotation opposite to the proper direction of rotation for a generator. Use a reduced voltage for large motors.

Brush Polarity

To maintain the original brush polarity when renewing the residual magnetism, the electrical polarities of the output leads and the exciting battery must be matched. In other words, the positive terminal of the battery must be connected to the positive output terminal of the generator and the negative battery terminal must be connected to the negative generator terminal.

The motorizing test should never be used for restoring residual flux if the generator armature is mechanically engaged to the prime mover and cannot rotate freely. A strong current through the motionless armature sets up a powerful magnetic field on the armature core. This magnetic field may overpower and reverse the main field flux, causing a reversal of the brush polarity when the generator is restarted. If there is any doubt as to whether or not the armature can be disconnected completely from the prime mover, it is preferable to isolate and energize the field circuit only, either by lifting the brushes or disconnecting the field leads.

CRITICAL FIELD RESISTANCE

A shunt generator may fail to reach its operating voltage even though its residual magnetic field is satisfactory. This failure may be due to excessive resistance in the field circuit. Any generator has *critical field resistance.* The presence of resistance in the field circuit in excess of this critical value causes the generator to fail to build up to its rated operating voltage.

Since field rheostats are used to control the voltage output at rated speed, it is important to reduce the resistance of the field rheostats to a minimum value before investigating other possible faults in the event of failure to develop rated voltage.

BRUSH CONTACT RESISTANCE

Contact resistance at the brushes is another reason for the failure of the generator to develop its operating voltage. Since the field circuit is *completed* through the armature, any resistance introduced at this point is effectively in the field circuit. Additional pressure applied to the brushes may indicate trouble from this source.

Improper connection of the field circuit leads at the brushes is also a cause of failure to build up rated voltage. An improper connection can be discovered by reversing these leads.

ROTATION

When a dc shunt generator is used in special applications, it may be necessary for the armature to rotate in a direction opposite to that specified by the manufacturer. To develop voltage buildup in these instances, the field circuit leads at the brushes must be reversed.

RATINGS

Shunt generators are rated for speed, voltage, and current. Generators used in aircraft and automobiles operate through a wide range of speeds, but must maintain a constant load voltage. Voltage regulators which automatically insert field resistance are used.

Generators designed for operation at a constant rated speed must not be operated above this value, unless the field circuit is protected from the effects of excessive current by current-limiting devices.

OUTPUT VOLTAGE CONTROL

Field rheostats are used to control the voltage output of shunt generators. At a given speed, the rheostat can be used only to bring the output voltage to values below the rated voltage obtainable without a field control. Values above the normal rated voltage can be obtained only by operating the generator above normal speed.

VOLTAGE REGULATION

The terms voltage regulation and voltage control are often confused. *Voltage control* refers to *intentional* changes in the terminal voltage made by manual or automatic regulating equipment, such as a field rheostat.

Voltage regulation refers to *automatic* changes in the terminal voltage due to reactions *within* the generator as the load current changes. For example, it is inherent in the design of a shunt generator for the output voltage to fall off as the load increases. If the drop is severe, the generator is said to have poor voltage regulation.

ACHIEVEMENT REVIEW

Select the correct answer for each of the following statements and place the corresponding letter in the space provided.

1. Most dc generators are   _____
    a. self-excited.
    b. excited by storage batteries.
    c. series wound.
    d. excited separately.

2. The field coils of a shunt generator are always connected   _____
    a. in parallel with a rheostat.
    b. in parallel with each other.
    c. in series with each other.
    d. across the line.

3. The voltage of a shunt generator is built up by   _____
    a. permanent magnetism.
    b. proper operation of the field rheostat.
    c. residual magnetism.
    d. increasing the speed.

4. The field windings of a shunt generator must have   _____

   a. full line current applied.
   b. comparatively low resistance.
   c. one ohm resistance per volt.
   d. comparatively high resistance.

5. Cutting resistance out of a shunt field circuit   _____

   a. cuts down the magnetic flux.
   b. decreases the terminal voltage.
   c. increases the load.
   d. increases the terminal output voltage.

6. Failure of a dc shunt generator to build up to its rated voltage
   can be due to   _____
   a. loss of residual magnetism.
   b. resistance greater than the critical field resistance.
   c. rotation of the armature in the direction opposite to that
       known to cause a voltage buildup.
   d. brush contact resistance effectively increasing the field cir-
       cuit resistance above the critical point.
   e. improper connection of the field circuit leads at the
       brushes.
   f. all of these.

7. Voltage control refers to a change that takes place   _____

   a. due to the operation of auxiliary regulating equipment.
   b. when the terminal voltage is increased.
   c. when the speed is regulated.
   d. automatically when the load is changed.

8. Voltage regulation refers to a change that takes place   _____

   a. when speed is regulated.
   b. when the terminal voltage is increased.
   c. automatically when the load is changed.
   d. when auxiliary equipment is used.

9. When the load is raised from minimum to maximum there is   _____

   a. no change in terminal voltage.
   b. an increase in terminal voltage.
   c. a decrease in terminal voltage.
   d. less change than in other generators.

10. Connect the following self-excited generator by drawing the proper connections in the terminal boxes.

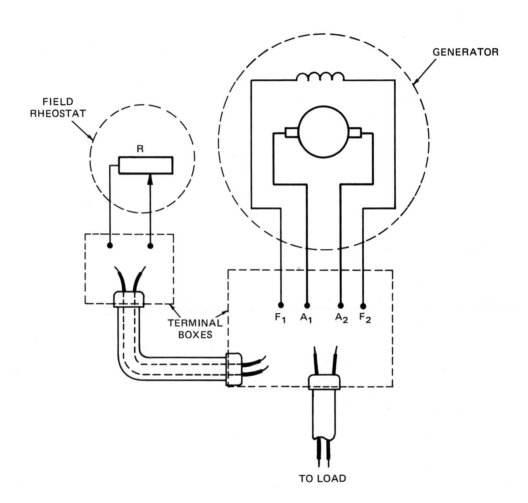

COMPOUND-WOUND DC GENERATOR

OBJECTIVES

After studying this unit, the student will be able to

- state the differences between a shunt generator and a compound-wound generator.
- define what is meant by a cumulative compound-wound generator and a differential compound-wound generator.
- describe how the voltage regulation of a generator is improved by the compound winding.
- list changes in output voltage at full load due to the effects of overcompounding, flat compounding, undercompounding, and differential compounding.
- draw the basic circuit.
- connect the generator.

The voltage regulation of a generator is an important factor in deciding the type of load to which the generator should be connected. For lighting loads, a constant terminal voltage should be maintained when the load current increases. A simple shunt generator can only do this if expensive regulating equipment is also used.

Generators designed to maintain a constant voltage within reasonable load limits may have a double winding in the field circuit (figure 4-1). The second winding is wound on top of, or adjacent to, the main winding. This second winding is called the *series winding* to distinguish it from the main shunt winding. The series winding has fewer turns than the shunt winding. Since the series winding is connected in series with the

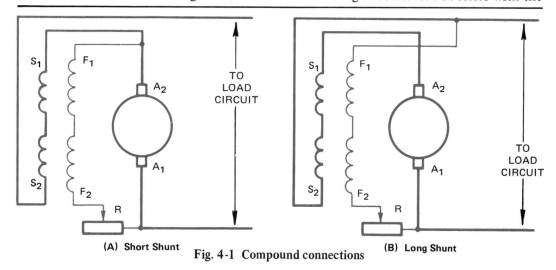

(A) Short Shunt   Fig. 4-1 Compound connections   **(B) Long Shunt**

armature and load, it carries the full-load current. A generator with such a *double-field winding* is called a *compound-wound generator.*

Figure 4-1 shows the basic circuits of two ways to connect a compound-wound generator: the long shunt and the short shunt. In the short shunt circuit (A), the main shunt field is connected directly across the brushes; in the long shunt circuit (B), the shunt field is connected across the combination of the armature and the series field. The operating characteristics of these circuits are quite similar, but the short shunt is preferred because of the simpler circuit.

COMPOUND FIELD WINDINGS

Two important details of the compound-wound generator must be considered: (1) the relative direction of the currents through both windings of a particular field pole, and (2) the magnetic effects which these currents can produce.

The series and shunt windings of a single pole of a compound-wound generator are shown in, figure 4-2. Winding (A) is the series winding through which the *load* current passes; winding (B) is the normal shunt winding. If the load current is in the direction shown in figure 4-2, the magnetizing force of the series winding (A) will aid the shunt winding (B) and increase the strength of the magnetic field. It is assumed that the current in the shunt winding is not strong enough to saturate the core. If the load current through the series winding is in the direction opposite to that shown in figure 4-2, its effect will be to weaken the magnetic field.

When the series winding is connected to aid the shunt winding, the generator is called a *cumulative compound-wound generator*; if the series winding is connected to oppose the magnetic field, it is called a *differential compound-wound generator.*

The action of two fields in changing the flux density can be used to improve the voltage regulation of a normal shunt generator. As a load is applied in the shunt

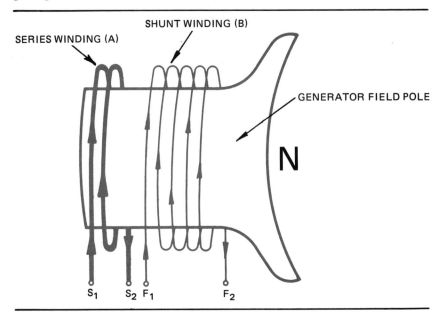

Fig. 4-2 Compound field windings

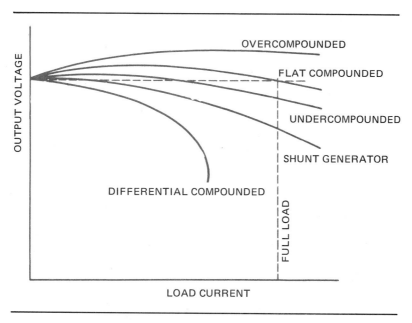

Fig. 4-3 Compound generator load characteristics

generator, the output voltage falls because of internal resistance, armature reaction, and the reduction of voltage applied to the field circuit. If the field strength can be *automatically increased* in *proportion* to load current as it increases, the output voltage can be maintained at a constant level, increased, or decreased. This is the effect which results due to the double winding in the compound generator. As the load current increases in a cumulative-compound connected generator, it passes through the series winding and increases the flux. The additional voltage induced by cutting this flux compensates for the voltage loss due to armature resistance, armature reaction, and lower shunt field voltage.

The number of turns in the series field helps determine the degree of compounding which is achieved. A large number of turns in the series winding produces *overcompounding* (a voltage increase at full load as compared to the output voltage at no load). A small number of series turns produces a reduced voltage at full load. This effect is called *undercompounding.*

Flat compound generators have the same voltage output at no load and full load. In industry, this type of generator is used where the distance between the generator and the load is short and line resistance is minimal. Overcompounding generators are used when the transmission distance is long, as in traction service, and the voltage at the end of the line must remain fairly constant.

A comparison of the voltage regulation of a shunt generator and a compound generator for both cumulative and differential connections is shown in figure 4-3.

OUTPUT VOLTAGE CONTROL

The rated voltage of a compound generator operating at rated speed is set by adjusting the field rheostat. Since the compounding effect of the series field changes with speed, it is important to operate a compound generator at its rated speed.

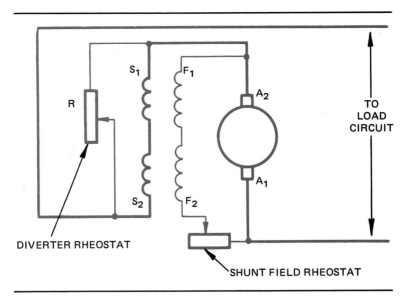

Fig. 4-4 Diverter circuit

Variation of Compounding

In general, compound-wound generators are designed by the manufacturer to have an overcompounding effect. The amount of compounding can be changed to any desired value by using a diverter rheostat across the series field. In figure 4-4, a *diverter rheostat* (R) is connected in shunt (parallel) with the series winding. If the resistance of the diverter is set at a high value, the load current passes through the series winding to produce a maximum compounding effect. If the diverter is set at its minimum value, no load current passes through the series winding and the generator acts like a normal shunt generator. By adjusting the rheostat to intermediate values, any degree of compounding within these limits can be obtained.

ACHIEVEMENT REVIEW

A.   Select the correct answer for each of the following statements and place the corresponding letter in the space provided.

1.   A compound-wound generator terminal connection box contains terminal leads   _____
     a. F_1, F_2, and A_1, A_2.
     b. S_1, S_2, and F_1, F_2.
     c. S_1, S_2, and A_1, A_2.
     d. S_1, S_2, F_1, F_2, and A_1, A_2.

2.   The series winding must be large enough to carry   _____
     a.   the total magnetic flux.
     b.   a 300% overload.
     c.   full line current.
     d.   full line voltage.

3. Select the type of generator that may be used for loads quite distant from the generator. _____
   a. overcompounded
   b. flat compounded
   c. undercompounded
   d. differential compounded

4. The normal voltage of a compound generator is changed by adjusting the _____
   a. series field shunt.
   b. brush setting.
   c. shunt field rheostat.
   d. equalizer.

5. The resistance of a series field diverter should be _____
   a. comparatively high.
   b. equal to the resistance of the series field.
   c. a variable resistor.
   d. comparatively low.

6. To achieve a maximum compounding effect, the diverter rheostat should be _____
   a. set at its minimum value.
   b. set at a high value.
   c. set at a value midway between its minimum and maximum values.
   d. removed from the series field circuit.

B. Select the correct answer to questions 7 to 12 from the following list and place it in the space provided.
   a. field poles
   b. diverter rheostat
   c. compound-wound generator
   d. saturate
   e. decrease
   f. shunt generator
   g. increase
   h. flat compounding
   i. shunt field rheostat
   j. overcompounding
   k. remain constant
   l. undercompounding

7. When it is necessary to provide automatic control of the voltage output at constant speed, the generator selected is a _____.

8. The current through the shunt winding of a compound generator is not sufficient to _____ the field poles.

9. The terminal voltage output of a cumulative compound-wound generator should _____ as the load current is increased.

10. When the output voltage of a generator is the same at both no load and full load, the generafor is called a _____ type.

11. Compound-wound generators are generally designed to be of the _____ type.

12. The amount of compounding which can be obtained from a generator is controlled by the _____.

THE DC SHUNT MOTOR

OBJECTIVES

After studying this unit, the student will be able to

- list the parts of a dc shunt motor.
- draw the connection diagrams for series shunt and compound motors.
- define torque and tell what factors affect the torque of a dc shunt motor.
- describe counter emf and its effects on current input.
- describe the effects of an increased load on armature current, torque, and speed of a dc shunt motor.
- list the speed control, torque, and speed regulation characteristics of a dc shunt motor.
- make dc motor connections.

The production of electrical energy, and its conversion to mechanical energy in electric motors of all types, is the basis of our industrial structure. Dc motor principles are given in ELECTRICITY 1.

CONSTRUCTION FEATURES

Dc motors closely resemble dc generators in construction features. In fact, it is difficult to identify them by appearance only. A motor has the same two main parts as a generator — the field structure and the armature assembly consisting of the armature core, armature winding, commutator, and brushes. Some general features of a dc motor are shown in figure 5-1A and B.

The Field Structure

The field structure of a motor has at least two pairs of field poles, although motors with four pairs of field poles are also used (figure 5-2A). A strong magnetic field is provided by the field windings of the individual field poles. The magnetic polarity of the field system is arranged so that the polarity of any particular field pole is opposite to that of the poles adjacent to it.

The Armature

The armature of a motor is a cylindrical iron structure mounted directly on the motor shaft (figure 5-2B). Armature windings are embedded in slots in the surface of the armature and terminate in segments of the commutator. Current is fed to these windings on the rotating armature by carbon brushes which press against the commutator

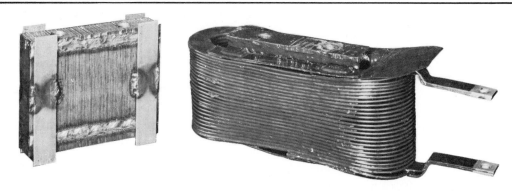

(A) Commutating-pole piece and commutating coil for shunt-wound, dc, 50-hp, 850-r/min, 230-V motor

(B) Main field coil and pole, and spring pads for shunt-wound, 50-hp, 850-r/min, 230-V motor

Fig. 5-1 Commutating pole and main field pole for a dc motor *(Courtesy General Electric Co.)*

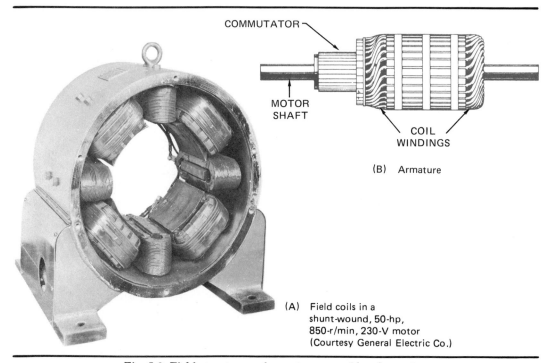

COMMUTATOR

MOTOR
SHAFT

COIL
WINDINGS

(B)   Armature

(A)   Field coils in a
       shunt-wound, 50-hp,
       850-r/min, 230-V motor
       (Courtesy General Electric Co.)

Fig. 5-2 Field structure and armature assembly of a motor

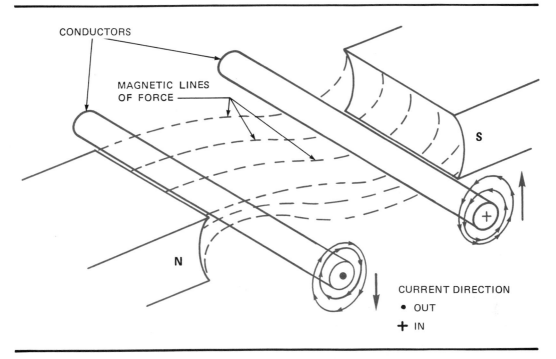

CONDUCTORS

MAGNETIC LINES
OF FORCE

S

N

CURRENT DIRECTION

• OUT

+ IN

Fig. 5-3 Torque, or force direction on a current-carrying conductor in a magnetic field

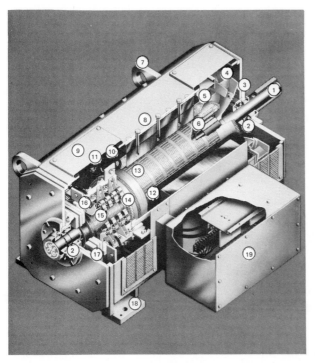

1. Large shaft
2. Bearings
3. Grease "meter"
4. Ventilating fan
5. Armature banding
6. Armature equalizer coil assembly
7. Lifting lugs
8. Frame
9. Inspection plate
10. Main field coil
11. Commutating coils
12. Main field coil
13. Armature
14. Commutator connections to armature turns
15. Commutator
16. Brushholder
17. Brushholder yoke
18. Mounting feet
19. Terminal conduit box

Fig. 5-4 Assembled 250-hp dc motor *(Courtesy of General Electric, DC Motor and Generator Department)*

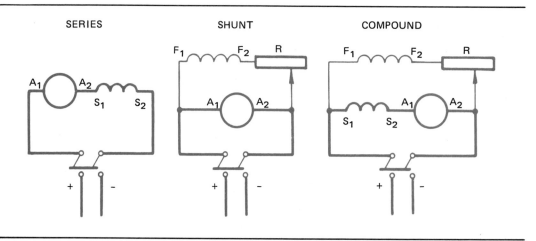

Fig. 5-5 Motor field connections

segments. This current in the armature sets up a magnetic field in the armature which acts against the magnetic field of the field poles to develop torque which causes the armature to turn (figure 5-3). The commutator changes the direction of the current in the armature conductors as they pass across poles of opposite magnetic polarity. Continuous rotation in one direction results from these reversals in the armature current.

Figure 5-4 is a cutaway view of a dc motor available with horsepower ratings ranging from 250 hp to 1,000 hp.

TYPES OF DC MOTORS

Shunt, series, and compound motors are all widely used. The schematic diagram for each type of motor is shown in figure 5-5. The selection of the type of motor to use is based on the mechanical requirements of the applied load. A shunt motor has the field circuit connected in shunt (parallel) with the armature, while a series motor has the armature and field circuits in series. A compound motor has both a shunt and a series field winding.

MOTOR RATINGS

Dc motors are rated by their voltage, current, speed, and horsepower output.

TORQUE

The rotating force at the motor shaft produced by the interaction of the magnetic fields of the armature and the field poles is called *torque*. The magnitude of the torque increases as the twisting force of the shaft increases. Torque is defined as the product of the force in pounds and the radius of the shaft or pulley in feet.

For example, a motor which produces a tangential force of 120 pounds at the surface of the shaft 2 inches in diameter has a torque of 10 foot-pounds (ft·lb).

$$\text{Torque} = \text{Force} \times \text{Radius}$$
$$= 120 \times 1/12 = 10 \text{ ft·lb}$$

Torque in a motor depends on the magnetic strengths of the field and the armature. Since the armature field depends on armature current, the torque increases as the armature current and hence the strength of the armature magnetic field increase.

It is necessary to distinguish between the torque developed by a motor when operating at its rated speed and the torque developed at the instant the motor starts. Certain types of motors have high torque at rated speed but poor starting torque. The many types of loads which can be applied to motors mean that the torque characteristic must be considered when selecting a motor for a particular installation.

STARTING CURRENT AND COUNTER ELECTROMOTIVE FORCE

The starting current of a dc motor is much higher than the current input while the motor is running freely at its rated speed. At the instant power is applied, the armature is motionless and the armature current is limited only by the very low armature circuit resistance. As the motor builds up to its rated speed, the current input decreases until the motor reaches its rated speed. At this point, the armature current stops decreasing and remains constant.

Factors other than armature resistance also limit the current. Figure 5-6 illustrates a demonstration which shows the "generator" action within a motor that accounts for the decrease in current with a speed increase.

In figure 5-6, a dc motor and a lamp (each with the same voltage rating) are connected in parallel to the dc source. A zero-center ammeter connected in the circuit indicates the amount and direction of the current to the motor. When the line switch is open (A), there is no current in any part of the circuit. When the switch is closed (B), the lamp lights instantly and the ammeter registers high current to the motor. The motor current decreases as the motor speed increases and remains constant when the motor reaches its rated speed. The instant the switch is opened, the ammeter deflection reverses. The lamp continues to light but grows dimmer as the motor speed falls.

Two conclusions can be made from this demonstration:

1. A dc motor develops an induced voltage while rotating.

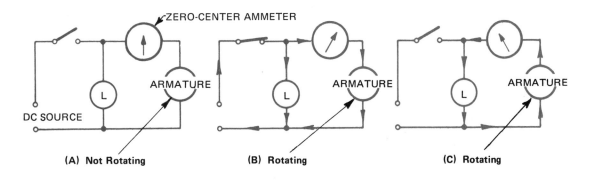

Fig. 5-6 Demonstration of counter emf

2. The direction of the induced voltage is opposite to that of the applied voltage and for this reason is called *counter emf.*

As the torque, or twisting effort, rotates the armature, the conductor coils of the armature cut the main field magnetic flux, as in a generator. This action induces a voltage into the armature windings which opposes line voltage.

The production of counter emf in a dc motor accounts for the changes in current to a motor armature at different speeds. When there is no current in the circuit, the motor armature is motionless and the counter emf is zero. The starting current is very high because only the ohmic resistance of the armature limits the current. As the armature starts to rotate, the counter emf increases and the current decreases. When the speed stops increasing, the value of the counter emf approaches the value of the applied voltage, but is never equal to it. The value of the voltage which actually forces current through the motor is equal to the difference between the applied voltage and the counter emf. At rated speed, this voltage differential will just maintain the motor at constant speed (figure 5-7A).

When a mechanical load is then applied to the motor shaft, both the speed and counter emf decrease. However, the voltage *differential increases* and causes an increase of input current to the motor. Any further increase in mechanical load produces a proportional increase in input current (figure 5-7B).

The increase in motor current due to an increase in mechanical load also can be explained in terms of the torque. Since torque depends upon the strength of the magnetic field of the armature which, in turn, depends upon the armature current, any increase in mechanical load must be accompanied by an increase in the armature current.

Since the starting current may be many times greater than the rated current under full load, large dc motors must not be connected directly to the power line for startup. The heavy current surges produce excessive line voltage drops which may damage the motor. The maximum branch-circuit fuse size for any dc motor is based on the full-load running current of the motor. Therefore, starters for dc motors generally limit the starting current to 150% of the full-load running current.

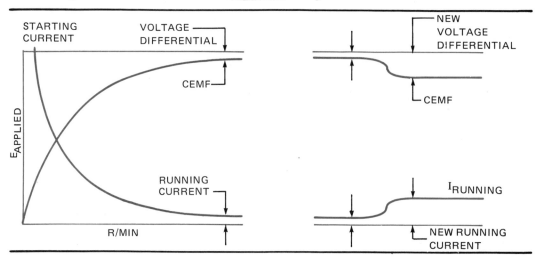

Fig. 5-7A Effects of counter-electromotive force on the armature current

Fig. 5-7B Effects of counter emf and $I_{armature}$ when the load is increased

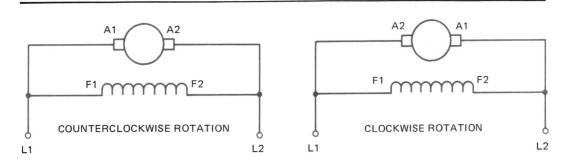

Fig. 5-8 Standard connections for shunt motors (From Herman/Alerich, *Industrial Motor Control*, copyright 1985 by Delmar Publishers Inc.)

ARMATURE REACTION

Armature reaction occurs in dc motors and produces a distortion of the field flux; this causes sparking at the brushes. Sparkless commutation is achieved in some motors by moving the brushes in the direction opposite to rotation. To counteract the effect of this field distortion, some motors are designed with field *interpoles* or *commutating poles* having windings which are connected in series with the armature circuit. This eliminates the need to shift brushes.

ROTATION

The direction of armature rotation of a dc motor depends on the direction of the current in the field circuit and the armature circuit (figure 5-8). To reverse the direction of rotation, the current direction in *either* the field or the armature must be reversed. Reversing the power leads does not reverse the direction of armature rotation because this situation causes *both* the field and armature currents to become reversed.

SPEED CONTROL AND SPEED REGULATION

The terms speed control and speed regulation are often used interchangeably, although their meanings are completely different. *Speed control* refers to changes in speed produced *intentionally* by the use of an auxiliary control, such as a field rheostat. *Speed regulation* refers to the changes in speed produced by changes *within* the motor due to a change in the load applied to the shaft. A motor that loses speed with a small increase in load is said to have poor speed regulation.

Speed Control

Dc motors are operated below normal speed by reducing the voltage applied to the armature circuit. Resistors connected in series with the armature may be used for voltage reduction. When the armature voltage is reduced while keeping the field current constant, the counter emf is too high. Therefore, the motor slows down to reduce the counter emf (figure 5-9A). The speed of a dc motor can also be brought below its rated speed by varying the voltage applied to the motor. However, this method is not used because there is a loss of torque along with the reduction in speed.

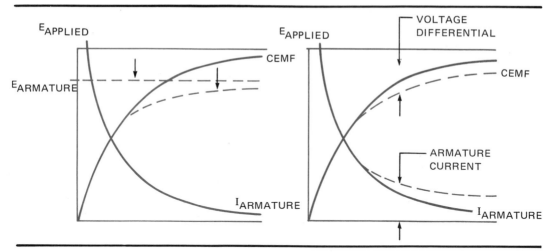

Fig. 5-9A To reduce speed, reduce the armature voltage while keeping the field current constant.

Fig. 5-9B To increase speed, reduce the field current while keeping the armature voltage constant.

A dc motor may be operated above its rated speed by reducing the strength of the field flux. A rheostat placed in the field circuit varies the field circuit resistance, the field current and, in turn, the field flux.

Although it seems reasonable that a reduction in field flux reduces the speed, the speed actually increases because the reduction of flux reduces the counter emf and permits the applied voltage to increase the armature current. The speed continues to increase until the increased torque is balanced by the opposing torque of the mechanical load. When the field flux is reduced while keeping the armature voltage constant, the counter emf in the armature drops. As a result, there is a larger voltage differential which causes an increase in armature current. This develops more torque to increase the speed of the motor (figure 5-9B).

Caution: Since motor speed increases with a decrease in field flux, the field circuit of a motor should never be opened when the motor is operating, particularly when it is running freely without a load. An open field may cause the motor to rotate at speeds that are dangerous to both the machine and to the personnel operating it. For this reason, some motors are protected against excessive speed by a field rheostat which has a no-field release feature. This device disconnects the motor from the power source if the field circuit opens.

THE SHUNT MOTOR

Two factors are important in the selection of a motor for a particular application: (1) the variation of the speed with a change in load, and (2) the variation of the torque with a change in load. A shunt motor is basically a constant speed device. If a load is applied, the motor *tends* to slow down. The slight loss in speed reduces the counter emf and results in an increase of the armature current. This action continues until the increased current produces enough torque to meet the demands of the increased load. As a result, the shunt motor is in a state of stable equilibrium because a change of load always produces a reaction that adapts the power input to the change in load.

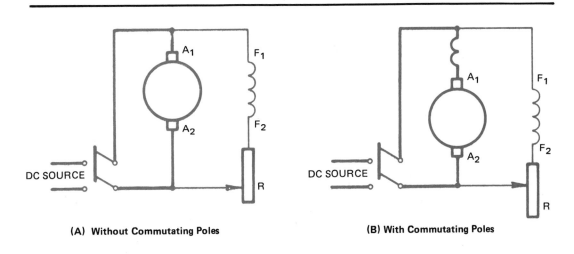

(A) Without Commutating Poles   (B) With Commutating Poles

Fig. 5-10 Shunt motor connections

The basic circuit for a shunt motor is shown in figure 5-10A. Note that only a shunt field winding is shown. Figure 5-10B shows the addition of a series winding to counteract the effects of armature reaction. From the standpoint of a schematic diagram, figure 5-10B represents a compound motor. However, this type of motor is not considered to be a compound motor because the commutating winding is not wound on the same pole as the field winding and the series field has only a few turns of wire in series with the armature circuit. As a result, the operating characteristics are those of a shunt motor. This is so noted on the nameplate of the motor by the terms compensated shunt motor or stabilized shunt motor.

Speed Control

A dc shunt motor has excellent speed control. To operate the motor above its rated speed, a field rheostat is used to reduce the field current and field flux. To operate below rated speed, resistors reduce the voltage applied to the armature circuit.

Rotation

The direction of armature rotation may be changed by reversing the direction of current in either the field circuit or the armature circuit. For a motor with a simple shunt field circuit, it may be easier to reverse the field circuit lead. If the motor has a series winding, or an interpole winding to counteract armature reaction, the same *relative* direction of current must be maintained in the shunt and series windings. For this reason, it is always easier to reverse the direction of the armature current.

Torque

A dc shunt motor has high torque at any speed. At startup, a dc shunt motor develops 150 percent of its rated torque if the resistors used in the starting mechanism are capable of withstanding the heating effects of the current. For very short periods of time, the motor can develop 350 percent of full load torque, if necessary.

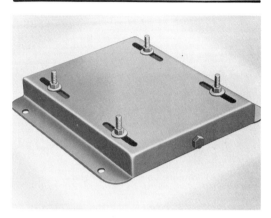

Fig. 5-11A Direct current motor, 1 hp to 5 hp *(Courtesy of General Electric, DC Motor and Generator Department)*

Fig. 5-11B Sliding base for dc motor *(Courtesy of General Electric, DC Motor and Generator Department)*

Speed Regulation

The speed regulation of a shunt motor drops from 5 percent to 10 percent from the no-load state to full load. As a result, a shunt motor is superior to the series dc motor, but is inferior to a differential compound-wound dc motor.

Figure 5-11A shows a dc motor with horsepower ratings ranging from 1 hp to 5 hp; figure 5-11B shows an optional mounting base. It is made of heavy gauge metal with slotted or sliding motor mounting bolts. This provides a convenient means of adjusting the belt drive tension.

ACHIEVEMENT REVIEW

A.   Select the correct answer for each of the following statements and place the corresponding letter in the space provided.

1. Dc motors are rated in _____
    a. voltage, frequency, current, and speed.
    b. voltage, current, speed, and torque.
    c. voltage, current, and horsepower.
    d. voltage, current, speed, and horsepower.

2. The generator effect in a motor produces a _____
    a. high power factor.
    b. high resistance.
    c. counter electromotive force.
    d. reduced line voltage.

3. A dc motor draws more current with a mechanical load applied to its shaft because the _____
    a. counter emf is reduced with the speed.
    b. voltage differential decreases.
    c. applied voltage decreases.
    d. torque depends on the magnetic strength.

4. The direction of rotation of a compound interpole motor may be reversed by reversing the direction of current flow through the _____
    a. armature.
    b. armature or field circuit.
    c. armature, interpoles, and series field.
    d. shunt field.

5. The speed of a dc motor may be reduced below its rated speed without losing torque by reducing the voltage at the _____
    a. motor.
    b. series field.
    c. armature.
    d. armature and field.

6. Advantages of dc motors are _____
    a. simplicity in construction.
    b. speed control above and below base speed.
    c. excellent torque and speed control.
    d. horsepower for size.

B. Complete the following statements.

7. The twisting force exerted on the shaft of a motor is called _____ and is due to the magnetic field interaction of the _____ and _____.

8. Field interpoles connected in series with the armature circuit of a motor help counteract the effects of _____ _____.

9. As a dc motor comes up to its rated speed, its armature current (decreases, remains the same, increases). (Underline the answer.)

10. The main factor controlling the armature current of a dc shunt motor operating at rated speed is the _____.

THE DC SERIES MOTOR

OBJECTIVES

After studying this unit, the student will be able to

- draw the basic connection circuit of a series dc motor.
- describe the effects on the torque and speed of a change in current.
- describe the effects of a reduction of a load on the speed of a dc series motor.
- connect a dc series motor.

Despite the wide use of alternating current for power generation and transmission, the dc series motor is often used as a starter motor in automobiles and aircraft. This type of motor is also used as a traction motor because of its ability to produce a high torque with only a moderate increase in power at reduced speed.

The basic circuit for the series motor is shown in figure 6-1. The field circuit has comparatively few turns of wire of a size that will permit it to carry the full-load current of the motor.

TORQUE

A series motor develops 500 percent of its full load torque at starting. Therefore, this type of motor is used for railway installations, cranes, and other applications for which the starting load is heavy.

It should be remembered that the shunt motor operates at constant speed. For this motor, any increase in torque requires a proportionate increase in armature current. In a series motor, the field is operated below saturation and any increase in load causes an increase of current in both the field and armature circuits. As a result, the armature flux and the field flux increase together. Since torque depends on the interaction of the armature and field fluxes, the torque increases as the square of the value of the current

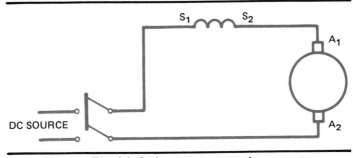

Fig. 6-1 Series motor connections

increases. Therefore, a series motor produces a greater torque than a shunt motor for the same increase in current. The series motor, however, shows a greater reduction in speed.

SPEED CONTROL AND SPEED REGULATION

The speed regulation of a series motor is inherently poorer than that of a shunt motor. If the mechanical load is reduced, a simultaneous reduction of current occurs in both the field and the armature. The reduction in the field current reduces the counter emf and the motor speeds up trying to rebuild the counter emf resulting from the reduced field flux. As a result, there is a greater increase in speed than would occur in a shunt motor for the same load change. If the mechanical load is removed entirely, the speed increases without limit and destruction of the armature through centrifugal force is certain to occur. For this reason, series motors are always permanently connected to their load.

If the maximum branch-circuit fuse size for any dc motor is limited to 150 percent of the full-load running current of the motor, the starters used with such motors must limit the starting current to 150 percent of the full-load current rating. Such starters must be equipped with an automatic, no-load release to prevent the armature from reaching dangerous speeds. The no-load release is set to open the circuit at the armature current corresponding to the maximum speed rating.

The speed of a series motor is controlled by varying the applied voltage. A series motor controller usually is designed to start, stop, reverse, and regulate the speed.

ROTATION

The direction of rotation may be reversed by changing the direction of the current either in the series field or the armature (figure 6-2).

MOTOR RATINGS

Series dc motors are rated for voltage, current, horsepower, and maximum speed.

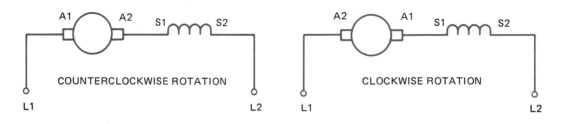

Fig. 6-2 **Standard connections for series motors** (From Herman/Alerich, *Industrial Motor Control*, copyright 1985 by Delmar Publishers Inc.)

ACHIEVEMENT REVIEW

Select the correct answer for each of the following statements and place the corresponding letter in the space provided.

1. The torque of a series motor   _____
    a. is lower in its starting value than the starting torque for a shunt motor of the same horsepower rating.
    b. depends on the flux of the armature only.
    c. increases directly as the square of the current increase.
    d. increases with a load increase, but causes less of a reduction in speed than a shunt motor for the same current increase.

2. For a series motor,   _____
    a. the field is operated below saturation.
    b. an increase in both the armature current and the field current occurs because of an increase in load.
    c. the reduction in speed due to an increase in load is greater than in the shunt motor.
    d. all of these.

3. Since a dc series motor has poor speed regulation,   _____
    a. a reduction in load causes an increase of current in both the field and armature.
    b. the removal of the mechanical load will cause the speed to increase without limit resulting in the destruction of the armature.
    c. it should not be connected permanently to its load.
    d. it does not require speed control.

4. The speed control for a dc series motor   _____
    a. is accomplished using a diverter rheostat across the series field.
    b. has an automatic no-field release feature included on all starters regardless of the limitations on the starting current.
    c. varies with the applied voltage.
    d. all of these.

5. A series motor controller usually is designed for   _____
    a. cranes.
    b. railway propulsion.
    c. starting loads when heavy.
    d. all of these.

6. Complete the electrical connections for the series motor.

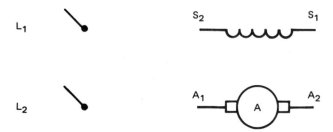

 # DC COMPOUND MOTORS

OBJECTIVES

After studying this unit, the student will be able to

- describe the torque, speed, rotation, and speed regulation and control characteristics of a cumulative compound-wound dc motor.

- perform the preliminary test for the proper installation of a cumulative compound motor.

- connect dc compound motors.

- describe the characteristics of a differential compound-wound dc motor.

- describe the characteristics of a cumulative compound-wound dc motor.

Compound-wound motors are used whenever it is necessary to obtain a speed regulation characteristic not obtainable with either a shunt or a series motor. Since many drives need a fairly high starting torque and a constant speed under load, the compound-wound motor is suitable for these applications. Some of the industrial applications include drives for passenger and freight elevators, stamping presses, rolling mills, and metal shears.

The compound motor has a normal shunt winding and a series winding on each field pole. As in the compound-wound dc generator, the series and shunt windings may be connected in long shunt (figure 7-1A), or short shunt (figure 7-1B).

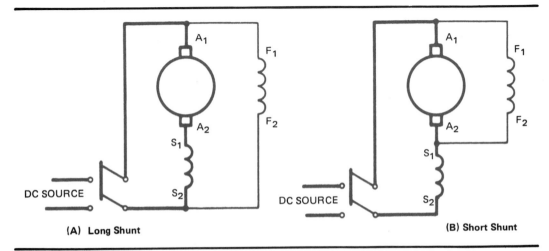

Fig. 7-1 Motor field connections

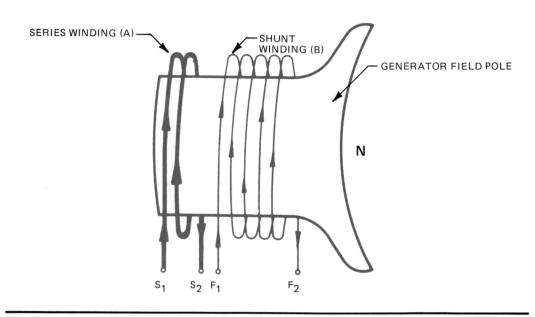

SERIES WINDING (A)

SHUNT WINDING (B)

GENERATOR FIELD POLE

N

S_1 S_2 F_1 F_2

Fig. 7-2 Compound field windings

When the series winding is connected to aid the shunt winding, the machine is a *cumulative compound motor.* When the series field opposes the shunt field, the machine is a *differential compound motor.* Using Fleming's right-hand rule for electromagnets, it can be seen that the two windings will either reinforce each other or try to cancel each other (figure 7-2).

TORQUE

The operating characteristics of a cumulative compound-wound motor are a combination of those of the series motor and the shunt motor. When a load is applied, the increasing current through the series winding increases the field flux. As a result, the torque for a given current is greater than it would be for a shunt motor. However, this flux increase causes the speed to decrease to a lower value than in a shunt motor. A cumulative compound-wound motor develops a high torque with any sudden increase of load. It is best suited for operating varying load machines such as punch presses.

SPEED

Unlike a series motor, the cumulative compound motor has a definite no-load speed and will not build up to destructive speeds if the load is removed.

Speed Control

The speed of a cumulative compound motor can be controlled by the use of resistors in the armature circuit to reduce the applied voltage. When the motor is to be used for installations where the rotation must be reversed frequently, such as in elevators, hoists, and railways, the controller used should have voltage dropping resistors and switching arrangements to accomplish reversal.

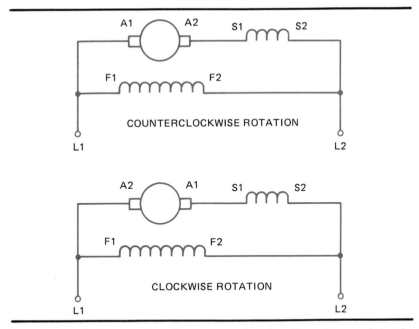

Fig. 7-3 Standard connections for compound motors (From Herman/Alerich, *Industrial Motor Control,* **copyright 1985 by Delmar Publishers Inc.)**

Speed Regulation

The speed regulation of a cumulative compound-wound motor is inferior to that of a shunt motor and superior to that of a series motor.

ROTATION

The rotation of a compound-wound motor can be reversed by changing the direction of the current in the field or the armature circuit, (figure 7-3). Since the series field coils must also be reversed if the shunt field is reversed, it is easier to reverse the current in the armature only.

PRELIMINARY TEST FOR CUMULATIVE COMPOUNDING

When a motor is first connected, it is important to determine the continuity of the shunt field circuit. In addition, for a compound-wound motor, the proper magnetic polarity of the shunt and series field must be determined. Standardized tests determine these conditions. For example, when the motor is connected to the controller and is ready for starting, disconnect the armature wire at the motor, close the line switch, and place the starter on the first contact point. Open the line switch slowly. If the field is intact there will be an arc at the switch. The absence of a spark indicates an open field circuit. This fault must be located and corrected before proceeding. A motor ordinarily will not start on an open field, but if it does start, it will race.

When the shunt field circuit tests complete, the motor should be started as a shunt machine. If the motor operates satisfactorily in the *desired* direction of rotation,

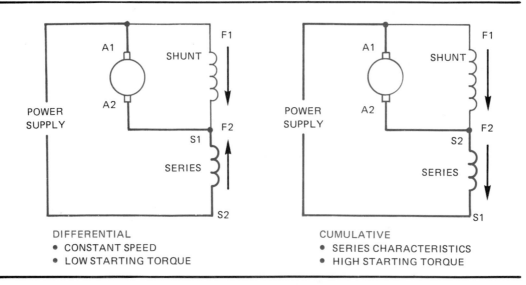

Fig. 7-4 Magnetic polarities of compound motors

shut the machine down. If it rotates in the opposite direction, shut it down and re-verse the shunt field leads. Restart the motor. If it now rotates in the desired direc-tion, shut it down.

Next, open the shunt field circuit, connect in the series field, and operate the machine momentarily as a series machine. As soon as the armature begins to turn, note the direction of rotation and shut down the machine.

If the armature rotates in the desired direction, connect in the shunt field circuit and the motor is ready for operation. If the direction of rotation as a series motor is opposite to the desired direction, reverse the series field leads and then connect the shunt field circuit. The motor is now ready for operation.

Differential Compounding

Excellent speed regulation can be obtained with a differential compound motor. When a motor is connected as a differential compound machine, the series field opposes the shunt field so that the field flux is decreased as a load is applied (figure 7-4). As a result, the speed remains substantially constant with an increase in load. With over-compounding, a slight increase in speed is possible with an increase in load. This speed characteristic is achieved only with a loss in the rate at which torque increases with load.

Since the field decreases with a load increase, a differential compound motor has a tendency to speed instability. When starting a differential motor, it is recommended that the series field be shorted since the great starting current in this field may over-balance the shunt field and cause the motor to start in the opposite direction.

A differential machine is connected and tested on installation, using the same procedure outlined for a cumulative compound motor. For the differential motor, however, the series windings should be connected in the opposite direction from that of the shunt winding.

Fig. 7-5 Steel ingot being moved by dc compound-wound motors *(Courtesy of General Electric, DC Motor and Generator Department)*

Figure 7-5 illustrates hot steel being moved precisely by dc compound-wound motors.

ACHIEVEMENT REVIEW

1. Circle the letter for each of the following statements which applies to a cumulative compound-wound dc motor.

   a. The speed regulation of a cumulative compound-wound dc motor is better than that of a shunt motor.

   b. The speed of a cumulative motor has a no-load limit.

   c. The speed of the motor decreases more for a given increase in load than does a differential motor.

   d. A cumulative motor has less torque than a shunt motor of the same hp rating for a given increase in armature current.

   e. The speed regulation of a cumulative motor is better than that of a series motor.

   f. A cumulative motor develops a high torque with a sudden increase of load.

   g. To reverse the direction of rotation, the current in either the armature or the shunt field must be reversed.

   h. A cumulative motor is connected so that the series flux aids the shunt winding flux.

   i. When installing a cumulative compound-wound motor, the direction of rotation should be the same when testing the motor for operation either as a series motor or a shunt motor.

2. Circle the letter for each of the following statements which applies to a differential compound-wound dc motor.

    a. A differential motor is used in applications where an essentially constant speed at various loads is required.

    b. The starting torque for a differential motor is higher than that of a cumulative motor.

    c. The motor may reverse its direction of rotation if started under a heavy load.

    d. This motor develops a speed instability since the flux field decreases with a load increase.

    e. When starting a differential motor, the shunt field should be shorted because of the great starting current.

3. Arrange the following steps numerically in the correct sequence to test for the proper connections to operate a cumulative compound-wound motor. Place the step number in the space provided, starting with number 1.

_____ a. Place starter on first contact point to test field.

_____ b. With motor shut down, open shunt field circuit.

_____ c. If rotation is in direction opposite to that desired, reverse the series field leads.

_____ d. Disconnect armature wire at motor and close line switch.

_____ e. An absence of spark indicates an open field circuit which must be located and corrected before proceeding with the test.

_____ f. If rotation is in direction opposite to that desired, shut down motor, reverse shunt field leads, and restart motor; rotation should be in desired direction.

_____ g. Slowly open line switch and observe for arc at switch indicating field is intact.

_____ h. Start motor as a shunt motor and observe rotation.

_____ i. Connect in series field, start motor and immediately shut it down while noting direction of rotation.

_____ j. Connect the shunt field circuit; the motor is now ready for operation.

SUMMARY REVIEW
OF UNITS 1-7

OBJECTIVE

- To give the student an opportunity to evaluate the knowledge and understanding acquired in the study of the previous seven units.

Select the correct answer for each of the following statements and place the corresponding letter in the space provided.

1. A generator series field diverter rheostat is always connected _____
   a. in parallel with the series field.
   b. in series with the series field.
   c. directly across the line.
   d. between the armature and the series field.

2. The voltage of an overcompounded generator _____
   a. decreases as the load increases.
   b. decreases as the load remains constant.
   c. increases as the load remains constant.
   d. increases as the load increases.

3. Brushes of a dc generator ride on the _____
   a. commutator.   c. shaft.
   b. armature.   d. commutating pole.

4. The field pole of a dc generator or motor is constructed with an iron core to _____
   a. decrease the magnetic flux.
   b. increase and concentrate the magnetic flux.
   c. decrease the residual magnetism.
   d. increase the eddy currents.

5. A generator with terminal markings A_1-A_2, F_1-F_2, S_1-S_2 is a _____
   a. separately-excited generator.   c. series generator.
   b. shunt generator.   d. compound generator.

6. The field core of a dc generator is _____
   a. the round part of the rotating field.
   b. wound with wire.
   c. usually round on large machines.
   d. the part of the generator that holds the armature in place.

7. In self-excited dc generators, initial field excitation is produced by   _____
    a. current in the coils.   c. magnetic flux.
    b. moving the field.   d. residual magnetism.

8. Dc generators and motors have   _____
    a. one pole.   c. pairs of poles.
    b. two poles.   d. four poles.

9. The twisting effect of a motor shaft is called its   _____
    a. turning power.   c. r/min.
    b. horsepower.   d. torque.

10. The twisting effect of a dc motor is produced primarily by   _____
    a. the armature.
    b. the rotor.
    c. a current-carrying conductor in a magnetic field.
    d. torque in the field coils.

11. A dc motor is required to maintain the same speed at no load
    and full load. This type of operation can only be obtained by
    using a   _____
    a. series motor.   c. differential compound-wound motor.
    b. shunt motor.   d. cumulative compound-wound motor.

12. As a load is applied to a dc shunt motor the   _____
    a. field current increases.
    b. counter emf increases.
    c. armature current increases.
    d. torque developed decreases.

13. The speed of a dc shunt motor   _____
    a. increases with an increase in load.
    b. decreases with an increase in applied voltage.
    c. decreases if the field strength is increased.
    d. decreases less than a series motor of the same hp for the
        same increase in load.

14. As load is applied to a dc series motor the   _____
    a. field current decreases.
    b. field voltage increases.
    c. armature current decreases.
    d. armature voltage increases.

15. The load requirements of a particular dc motor installation
    require extremely high starting torque. If speed regulation is
    not important, use a   _____
    a. series motor.
    b. shunt motor.
    c. differential compound-wound motor.
    d. cumulative compound-wound motor.

16. As a load is applied to a cumulative compound-wound dc
    motor its   _____
    a. speed decreases.
    b. counter emf decreases.
    c. torque decreases.
    d. series field current decreases.

17. In a cumulative compound-wound dc motor, the   _____
    a. series winding develops the major part of the total flux.
    b. series and shunt windings develop field flux in the same
       direction.
    c. shunt winding must be connected across the brushes.
    d. series windings do not pass the shunt field current.

18. In a dc shunt motor *all but one* of the following are true. _____
    a. Torque is proportional to the field current.
    b. The same voltage is applied to armature and field circuits.
    c. The no-load speed is controlled by the impressed voltage.
    d. The motor is suitable for installations requiring substan-
       tially constant speed with variable loading.

19. In a differential compound-wound dc motor   _____
    a. the series and shunt fields establish flux in the same
       direction.
    b. the series winding acts to reduce speed as load is applied.
    c. an increase in total current input as the result of loading
       increases the shunt field current.
    d. changes in torque result in change of current in the series
       field windings.

20. If the direction of field flux and the direction of armature
    current are changed, the torque developed by the motor is   _____
    a. stronger.   c. the same.
    b. less.   d. reversed.

21. A generator shunt field winding is   _____
    a. high resistance.   c. noninductively wound.
    b. low resistance.   d. embedded in the armature.

22. For proper operation in a four-lead dc motor, leads S_1 and
    S_2 should be connected to A_1 and A_2 in   _____
    a. parallel.   c. shunt.
    b. series.   d. series-parallel.

23. Decreasing the resistance of a generator field rheostat   _____
    a. decreases the flux.   c. increases the voltage.
    b. decreases the voltage. d. decreases the speed.

24. A series field, if connected across a motor armature and energized, _____
    a. makes the motor race dangerously.
    b. causes a short circuit.
    c. creates excessive flux.
    d. acts as a shunt field.

MANUAL STARTING RHEOSTATS FOR DC MOTORS

OBJECTIVES

After studying this unit, the student will be able to

- list the primary functions of the three-terminal starting rheostat and the four-terminal starting rheostat.
- demonstrate the proper connection and startup procedure for a three-terminal starting rheostat on a shunt motor and on a cumulative compound-wound motor.
- demonstrate the proper connection and startup procedure for a four-terminal starting rheostat on a shunt motor and on a cumulative compound-wound motor.
- state basic dc motor starting principles common to other motor starters.

Two factors limit the current taken by a motor armature from a direct-current source: (1) the counter electromotive force, and (2) the armature resistance. **Since there is no counter emf when the armature is at a standstill, the current taken by the armature will be abnormally high. As a result, the armature current must be limited by an external resistor, such as a starting rheostat.** *Electric Motor Control* covers in detail this method of limiting armature current.

A *starting rheostat* or motor starter is described by the National Electrical Manufacturers' Association as a device designed to accelerate a motor to its normal rated speed in one direction of rotation. In addition, a motor starter limits the current in the armature circuit to a safe value during the starting or accelerating period. Two common types of manual starting rheostats are:

- three-terminal starting rheostat, and
- four-terminal starting rheostat.

The electrical technician is expected to know how each type of starting rheostat is connected to direct-current shunt and compound motors. The technician should also know the specific applications and limitations of each type of motor starter.

THREE-TERMINAL STARTING RHEOSTAT

The three-terminal starting rheostat has a tapped resistor enclosed in a ventilated box. Contact buttons located on a panel mounted on the front of the box are connected to the tapped resistor. A movable arm with a spring reset can be moved over the contact buttons to cut out sections of the tapped resistor.

The connection diagram for a typical three-terminal starting rheostat is shown in figure 9-1. Note that the starter has three terminals or connection points and that it is connected to a shunt motor.

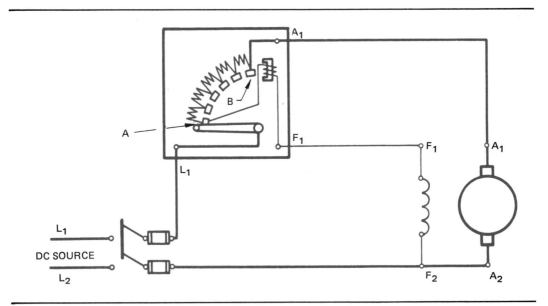

Fig. 9-1 Connections for a three-terminal starting rheostat

When the arm of the rheostat is moved to the first contact, A, the armature (which is in series with the starting resistance) is connected across the source. The shunt field, in series with the holding coil, is also connected across the source. The initial current inrush to the armature is limited to a safe value by the starting resistance. In addition, the shunt field current is at a maximum value and provides a good starting torque.

As the arm is moved to the right toward contact B, the starting resistance is reduced and the motor accelerates to its rated speed. When the arm reaches contact B, the armature is connected directly across the source voltage, and the motor will have attained full speed.

The holding coil is connected in series with the shunt field and provides a no-field release. **If the shunt field opens, the motor speed will become dangerously high if the armature circuit remains connected across the source.** Therefore, in the event of an open-circuited shunt field, the holding coil of the starting rheostat becomes demagnetized and the arm returns to the off position.

Note that the starting resistance is in series with the shunt field when the arm is in the run position at contact B. This additional resistance has practically no effect on the speed as the starting resistance is small when compared with the shunt field resistance.

To operate a three-terminal starter, first close the line switch. **Then, move the starting arm from the off position to contact A. Continue to move the arm slowly toward contact B, pausing on each intermediate contact for a period of one to two seconds.** By moving the arm slowly toward the run position at contact B, the motor will accelerate uniformly to its rated speed without an excessive inrush of current to the armature. However, do not hold the arm on any one contact between A and B for too long a period of time. These starting resistors are designed to carry the starting current for a short period of time only. In other words, do not control the speed of the motor by holding the arm for any length of time on any contact between A and B.

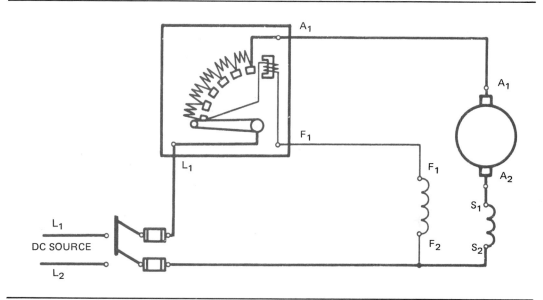

Fig. 9-2 Three-terminal starting rheostat connected to a cumulative compound-wound motor

If it is necessary to control the speed of the motor, **do not use a three-terminal starter.** Since the current in the shunt field and holding coil may be reduced to a value insufficient to hold the arm against the action of the reset spring, the reset spring will return the arm to the off position. Thus, the motor will become disconnected from the current source.

When a motor is to be disconnected from the current source, first open the line switch quickly. Then, check to see that the spring reset returns the starting arm to the off position.

Figure 9-2 shows the connections for a three-terminal starting rheostat used with a cumulative compound-wound motor. Note that these connections are almost the same as those of a three-terminal starting rheostat connected to a shunt motor; the only change in figure 9-2 is the addition to the motor of the series field.

FOUR-TERMINAL STARTING RHEOSTAT

A four-terminal starting rheostat performs the same functions as a three-terminal starting rheostat:

- it accelerates a motor to rated speed in one direction of rotation.
- it limits the starting surge of current in the armature circuit to a safe value.

In addition, the four-terminal starting rheostat may be used where a wide range of motor speeds is necessary. A field rheostat may be inserted in series with the shunt field circuit to obtain the desired speed.

Figure 9-3 shows a four-terminal starting rheostat. Note that the holding coil is not connected in series with the shunt field as it is in the three-terminal starting rheostat. The holding coil in figure 9-3 is connected in series with a resistor across the

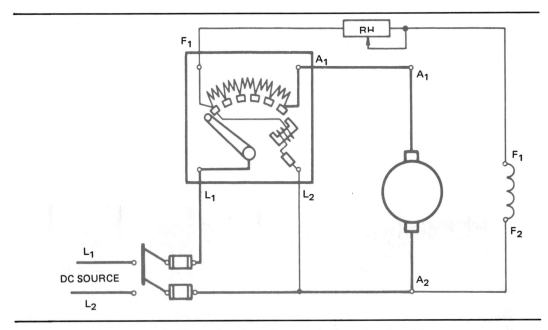

Fig. 9-3 Connections for a four-terminal starting rheostat

source. Note also that the holding coil circuit is connected across the source and, as a result, four terminal connection points are necessary.

The holding coil of the rheostat is connected across the source and acts as a no-voltage release. For example, if the line voltage drops below the desired value, the attraction of the holding coil is decreased, and the reset spring will then return the arm to the off position.

When a four-terminal starting rheostat is used, the speed of the motor is controlled by varying the resistance of the field rheostat connected in series with the shunt field circuit. The speed is increased above the rated speed by inserting resistance in the field rheostat.

When a motor using a four-terminal starting rheostat is to be disconnected from the source, **first cut out all resistance in the field rheostat. Then, open the line switch and check to see that the spring reset returns the starting arm to the off position.**

By removing all resistance from the field rheostat, the strength of the shunt field is increased. Thus, when the motor is restarted, it will have a strong field and strong starting torque.

Figure 9-4 shows a four-terminal starting rheostat connected to a cumulative compound-wound motor. Note the similarity in connections for a shunt motor (figure 9-3) and a compound-wound motor (figure 9-4). The only change in figure 9-4 is the addition of the series field.

NATIONAL ELECTRICAL CODE RULES FOR MOTOR STARTERS

The National Electrical Code states that a motor starter shall be marked with voltage and horsepower ratings, and the manufacturer's name and identification symbols, such as style or type numbers.

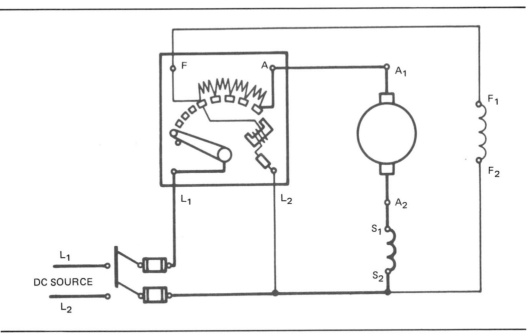

Fig. 9-4 Four-terminal starting rheostat connected to a cumulative compound-wound motor

Electrical codes require that the horsepower rating of a starter must not be smaller than the horsepower rating of the motor. In addition, the fuse protection for dc motors must be no greater than a percentage of the full-load current rating of the motor. Therefore, the motor starter must limit the starting current to a value which is no greater than a percentage (specified by the electrical code) of the full-load current rating of the motor.

It is recommended at this time to review the National Electrical Code section on motors, motor circuits, and controllers.

SAFETY PRECAUTIONS FOR DC MOTOR STARTERS

Current must be limited when starting the armature. A no-field release is provided to prevent runaway acceleration.

To operate a three- or four-terminal starter, the starting arm is moved slowly, pausing for one or two seconds only before moving to the next position.

Three-Terminal Starter

To decelerate the motor, first open the line switch. Check the starting arm and return it to the OFF position by using the reset spring.

Four-Terminal Starting Rheostat

Decelerate the motor by cutting out all resistance in the field rheostat. Open the line switch. Check the starting arm and return it to the OFF position by using the reset spring.

ACHIEVEMENT REVIEW

1. What are the two functions of a motor starter?

    a. _____

    b. _____

2. Show the connections of a three-terminal starting rheostat to a shunt motor.

3. State one advantage of a three-terminal starting rheostat.

    _____

    _____

    _____

    _____

4. Name one limitation of a three-terminal starting rheostat.

    _____

    _____

    _____

    _____

5. Complete the connections in the following figure to show the shunt motor properly connected to the three-terminal starting rheostat.

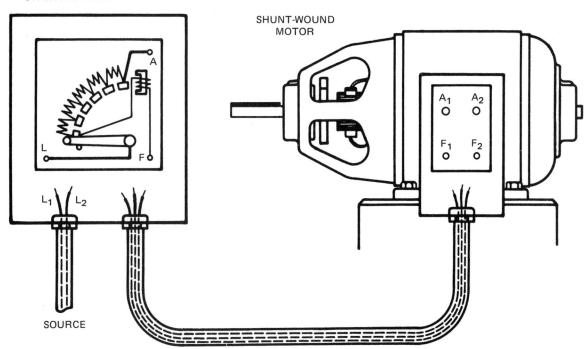

6. Show the connections of a four-terminal starting rheostat to a shunt motor.

7. What is one advantage of a four-terminal starting rheostat?

_____

_____

_____

_____

8. List the items that should be marked on the nameplate of a motor starter to comply with National Electrical Code requirements.

_____

_____

_____

_____

9. Complete the connections in the following figure to show that the cumulative compound-wound motor can be started from the four-terminal starting rheostat. Also connect the field rheostat in the circuit for above-normal speed control.

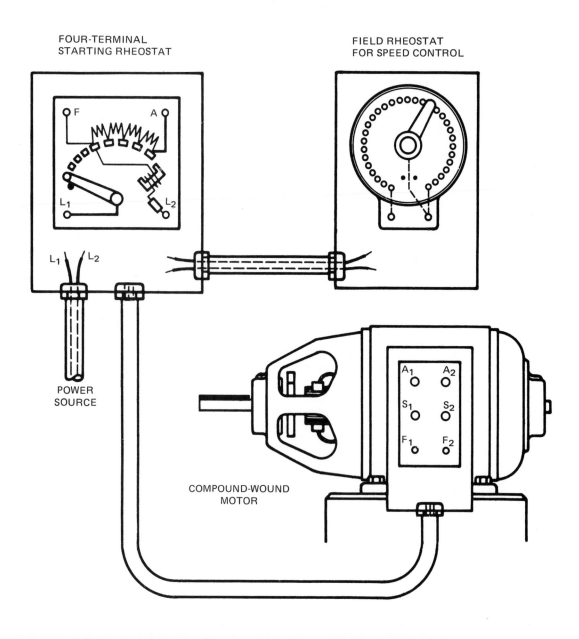

FOUR-TERMINAL
STARTING RHEOSTAT

FIELD RHEOSTAT
FOR SPEED CONTROL

POWER
SOURCE

COMPOUND-WOUND
MOTOR

10. What is the full-load current rating of a 5-hp, 240-V motor? (Refer to the National Electrical Code, if necessary.)   _____

11. What size conduit is required between the 5-hp motor and the starting box, using type THHN wire?   _____

SPECIAL DC STARTING RHEOSTATS AND CONTROLLERS

OBJECTIVES

After studying this unit, the student will be able to

- describe the operation of a series motor starter with no-voltage protection.
- describe the operation of a series motor starter with no-load protection.
- describe the actions occurring at each forward and reverse position of a drum controller.

Series motors require a special type of starting rheostat called a *series motor starter.* These starting rheostats serve the same purpose as the three- and four-terminal starting rheostats used with shunt and compound motors. However, the internal and external connections for the series motor starter differ from the connections of the other types of starting rheostats.

Series and cumulative compound motors are often used for special industrial applications which require provisions for reversing the direction of rotation and varying the speed of the motor. A manually operated controller, called a *drum controller*, may be used for these applications. The operation of drum controllers is discussed later in this unit.

STARTING RHEOSTATS FOR DC SERIES MOTORS

Series motor starting rheostats are of two types: one type of starter has no-voltage protection, and the other type has no-load protection.

Starter with No-Voltage Protection

A series motor starter with no-voltage protection is shown in figure 10-1. The holding coil circuit of this starter is connected across the source voltage. There is no shunt field connection on this type of starter as it is used only with series motors. This type of starter is used to accelerate the motor to rated speed. In the event of voltage failure, the holding coil no longer acts as an electromagnet. The spring reset then quickly returns the arm to the off position. Thus, the motor is protected from possible damage due to low-voltage conditions.

To disconnect a motor using this type of starting rheostat, open the line switch. Check to be sure that the arm returns to the off position.

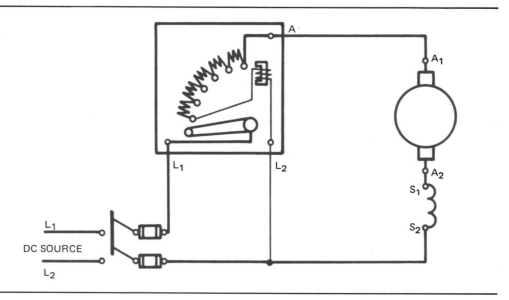

Fig. 10-1 Series motor starter with no-voltage protection

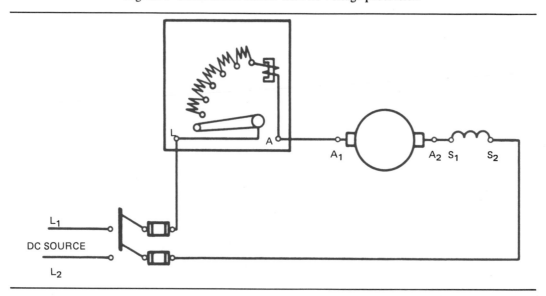

Fig. 10-2 Series motor starter with no-load protection

Starter with No-Load Protection

A series motor starter with no-load protection is shown in figure 10-2. The holding coil is in series with the armature circuit. Because of the relatively large current in the armature circuit, the holding coil consists of only a few turns of heavy wire. Note in the figure that separate terminal connections for the shunt field and holding coil are not provided. There are only two terminals — one marked L (line) and one marked A (armature).

The same care is required in starting a motor with this type of starting rheostat as is required with three- and four-terminal starting rheostats. The arm is slowly moved from the off position to the run position, pausing on each contact button for a period of one to two seconds. The arm is held against the tension of the reset spring by the holding coil connected in series with the armature. If the load current to the motor drops to a low value, the holding coil weakens and the reset spring returns the arm to the off position. This is an important protective feature. **Recall that a series motor may reach a dangerously high speed at light loads.** Therefore, if the motor current drops to such a low value that the speed becomes dangerous, the holding coil will release the arm to

100-ampere cam drum switch used for crane hoist

Fig. 10-3 Drum controller *(Courtesy of General Electric Company)*

the off position. In this way, it is possible to avoid damage to the motor due to excessive speeds.

To stop a series motor connected to this type of starting rheostat, open the line switch. Check to be sure that the arm returns to the off position.

DRUM CONTROLLERS

Drum controllers are used when an operator is controlling the motor directly. The drum controller is used to start, stop, reverse, and vary the speed of a motor. This type of controller is used on crane motors, elevators, machine tools, and other applications in heavy industry. As a result, the drum controller must be more rugged than the starting rheostat.

A drum controller with its cover removed is shown in figure 10-3. The switch consists of a series of contacts mounted on a movable cylinder. The contacts, which are insulated from the cylinder and from one another, are called movable contacts. There is another set of contacts, called stationary contacts, located inside the controller. These stationary contacts are arranged to touch the movable contacts as the cylinder is rotated. A handle, keyed to the shaft for the movable cylinder and contacts, is located on top of the drum controller. This handle can be moved either clockwise or counterclockwise to give a range of speed control in either direction of rotation. The handle can remain stationary in either the forward or reverse direction due to a roller and a notched wheel. A spring forces the roller into one of the notches at each successive position of the controller handle to keep the cylinder and movable contacts stationary until the handle is moved by the operator.

A drum controller with two steps of resistance is shown in figure 10-4. The contacts are represented in a flat position in this schematic diagram to make it easier to trace the circuit connections. To operate the motor in the forward direction, the set of contacts on the right must make contact with the center stationary contacts. Operation in the reverse direction requires that the set of movable contacts on the left makes contact with the center stationary contacts.

Note in figure 10-4 that there are three forward positions and three reverse positions to which the controller handle can be set. In the first forward position, all of the resistance is in series with the armature. The circuit path for the first forward position is as follows:

1. Movable fingers a, b, c, and d contact the stationary contacts 7, 5, 4, and 3.

2. The current path is from the positive side of the line to contact 7, from 7 to a, from a to b, from b to 5, and then to armature terminal A_1.

3. After passing through the armature winding to terminal A_2 the current path is to stationary contact 6, and then to stationary contact 4.

4. From contact 4 the current path is to contact c, to d, and then to contact 3.

5. The current path then goes through the armature resistor, to the series field, and then back to the negative side of the line.

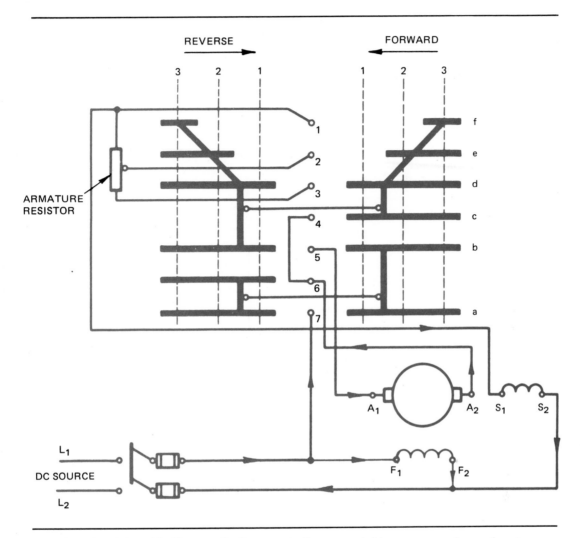

Fig. 10-4 Schematic diagram of a drum controller connected to a compound-wound motor

The shunt field of the compound motor is connected across the source voltage. On the second forward position of the controller handle, part of the resistance is cut out. The third forward position cuts out all of the resistance and puts the armature circuit directly across the source voltage.

In the first reverse position, all of the resistance is inserted in series with the armature.

Figure 10-5 shows the first position of the controller in the reverse direction. The current in the armature circuit is reversed. However, the current direction in the shunt and series fields is the same as the direction for the forward positions. Remember that an earlier unit showed that a change in current direction in the armature only resulted in a change in the direction of rotation.

The second reverse position cuts out part of the resistance circuit. The third reverse position cuts out all of the resistance and puts the armature circuit directly across the source. Drum controllers with more positions for a greater control of speed can be

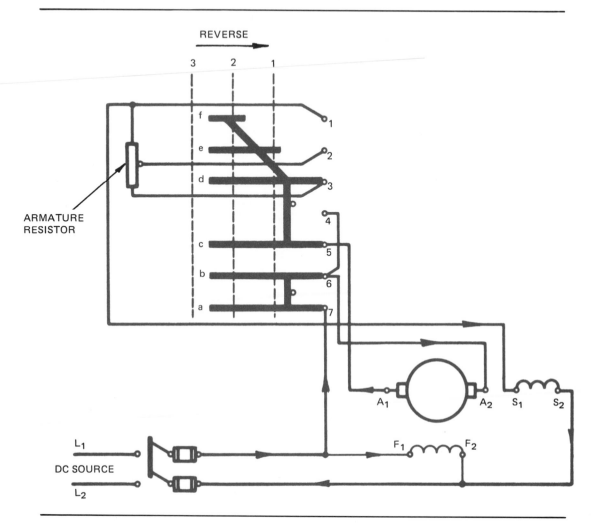

Fig. 10-5 First position of controller for reverse direction

obtained. However, these controllers all use the same type of circuit arrangement shown in this unit.

ACHIEVEMENT REVIEW

1. Show the internal connections of a series starting rheostat with no-voltage protection.

2. Show the internal connections of a series starting rheostat with no-load protection.

3. Show the circuit connections for a series motor used with a series starting rheostat with no-load protection.

4. Show the circuit connections for a series motor used with a series starting rheostat with no-voltage protection.

5. Why is a drum controller used in many industrial applications?

_____

_____

Complete the following statements.

6. In a series starter with no-voltage protection, the holding coil is connected across the _____ .

7. A series starter with no-load protection is used to prevent the series motor from reaching _____ at low loads.

8. A drum controller gives the following types of control for a direct-current motor:

_____

_____ .

BASIC PRINCIPLES OF AUTOMATIC MOTOR CONTROL

UNIT 11

OBJECTIVES

After studying this unit, the student will be able to

- list several factors to be considered when selecting and installing electric motor control equipment.
- explain the purpose of a contactor.
- describe the basic operation of a contactor and relay.
- list the steps in the operation of a control circuit using start and stop pushbuttons.
- interpret simple automatic control diagrams.
- draw a simple magnetic control circuit.

Motor control was a simple problem when motors were used to drive a common line shaft to which several machines were connected. In this arrangement, it was necessary to start and stop only a few times daily.

With individual drive, however, the motor is an integral part of the machine and the motor controller must be designed to meet the needs of the machine to which it is connected.

As a result, the modern motor controller does not just start, stop, and control the speed of a motor. The controller may also be required to sense a number of conditions, including changes in temperature, open circuits, current limitations, overload, smoke density, level of liquids, or the position of devices. Manual control is limited to pressing a button to start or stop the entire sequence of operations at the machine or from a remote position.

The electrician must know the symbols and terms used in automatic control diagrams to be able to wire, install, troubleshoot, and maintain automatic control equipment.

CLASSIFICATION OF AUTOMATIC CONTROLLERS

Purpose

Factors to be considered in selecting motor controllers include starting, stopping, reversing, running, speed and sequence control, and protection.

Operation

The motor may be controlled manually by an operator using a switch or a drum controller. Remote control uses contactors, relays, and pushbuttons.

Fig. 11-1A Dc magnetic relay *(Courtesy of General Electric Company)*

Fig. 11-1B Dc operated control relay *(Courtesy of Square D Company)*

CONTACTORS

Contactors, or relays (figure 11-1) are required in automatic controls to transmit varying conditions in one circuit to influence the operation of other devices in the same or another electrical circuit. Relays have been designed to respond to one or more of the following conditions:

voltage	overvoltage	undervoltage
current	overcurrent	undercurrent
current direction	differential current	power (watts)
power direction	volt-amperes	frequency
phase angle	power factor	phase rotation
phase failure	impedance	speed
	temperature	

Magnetic switches are widely used in controllers because they can be used with remote control, and are economical and safe.

A relay or contactor usually has a coil which can be energized to close or open contacts in an electrical circuit. The coil and contacts of a relay are represented by symbols on the circuit diagram or schematic of a controller. Symbols commonly used to represent contactor elements are shown in figure 11-2.

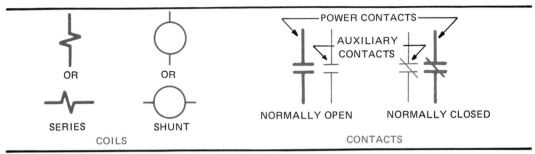

Fig. 11-2 Schematic symbols for contactor elements

If the control coil is connected in series in the motor power circuit, the heavy line symbol shown at the left of figure 11-2 is used. If the coil is connected in parallel (shunt), the light line symbol is used.

A series coil has a large cross-sectional conductor area with few turns; such a coil will carry large currents. A shunt coil has a small wire size with many turns; it carries small currents. It is possible for a series coil and a shunt coil to have the same ampere-turns, resulting in similar magnetic results.

Contacts which are open when the coil is deenergized are known as *normally open* contacts and are indicated by two short parallel lines. Contacts which are closed when the coil is deenergized are called *normally closed* contacts and are indicated by a slant line drawn across the parallel lines.

To minimize heavy arcing which burns the contacts, a dc contactor usually is equipped with a *blowout coil* and an *arc chute.* Figure 11-3 shows a magnetic contactor which is provided with a blowout coil and an arc chute.

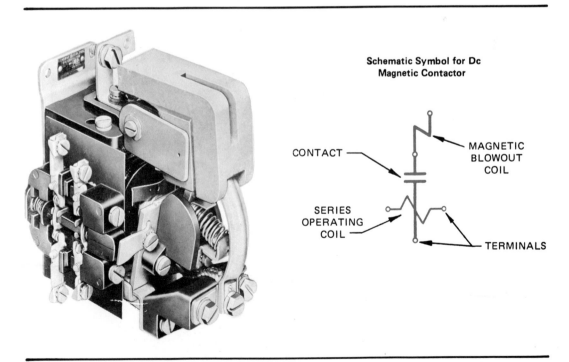

Fig. 11-3 Dc magnetic contactor with blowout coil and arc chute *(Courtesy of General Electric Company)*

Fig. 11-4 Behavior of arc with correctly designed blowout

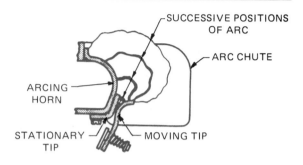

A. Pushbutton Control Station

B. Multiple Control Station

Fig. 11-5 Pushbutton stations *(Courtesy of Square D Company)*

Fig. 11-6 Symbols for pushbutton contacts

When a heavy current is broken by the contacts of the contactor, an arc occurs. Figure 11-4 illustrates the behavior of the arc as it is quickly extinguished by the electromagnetic and thermal action of the magnetic blowout coil and arc chute.

PUSHBUTTONS

Pushbutton stations (figures 11-5A and B), are spring-controlled switches and, when pushed, are used to complete motor or motor control circuits. Figure 11-5B shows multiple control stations, with pushbuttons, selector switches, and pilot indicating lights. Note the "mushroom" stop button for easy access. This is for convenience and safety.

The symbols used in schematic drawings to represent pushbutton contacts are given in figure 11-6.

TYPICAL CONTROL CIRCUIT

Figure 11-7 is an elementary control circuit with start and stop buttons and a sealing circuit. The following sequence describes the operation of the circuit.

1. When the start button is pressed to close contacts 2–3, current flows from L_1 through normally closed contacts 1–2 of the stop button, through contacts 2–3 of the normally open contacts of the start button, and through coil M to Line L_2.

2. The current in coil M causes the contact M to close. Thus, the sealing circuit around contacts 2–3 of the start button closes. The start button may now be released, and even though the spring of the pushbutton opens contacts 2–3, coil M remains energized and holds contacts M closed to maintain a sealing circuit around the normally open contacts 2–3 of the start button. Coil

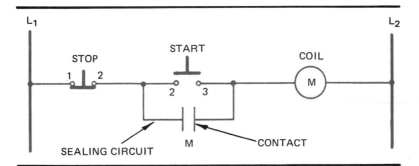

Fig. 11-7 A control circuit with start and stop buttons and sealing circuit

M, being energized, also closes M contacts in the power circuit to the motor (not shown).

3. If the stop button is momentarily pressed, the circuit is interrupted at contacts 1-2 and coil M is deenergized. Contacts M then open and coil M cannot be energized until the start button again closes contacts 2-3.

ACHIEVEMENT REVIEW

Select the correct answer for each of the following statements and place the corresponding letter in the space provided.

1. Early motor installations consisted of   _____
   a. individual drives.
   b. a common line shaft drive.
   c. automatically controlled motor drives.
   d. remotely controlled motors.

2. Individual motor drives require   _____
   a. single-phase motors.
   b. automatic controllers.
   c. speed rheostats.
   d. gear heads.

3. Automatic dc motor controllers are designed to respond to changes in temperature, open circuits, current limitations, and   _____
   a. wire size.
   b. fuse rating.
   c. speed acceleration.
   d. brush assembly.

4. Interpretation of automatic control circuits requires the recognition of   _____
   a. color.   c. ratings.
   b. electrical circuit symbols. d. parallel circuits.

5. A relay symbol shows the   _____
   a. number of turns in a coil.
   b. relay current rating.
   c. relative position of the component parts.
   d. size of the contacts.

6. A relay is classified as a piece of electrical equipment with at least one   _____
   a. coil.
   b. resistor.
   c. coil operating one contact.
   d. coil operating two contacts.

7. Normally open contacts are _____
   a. open at all times.
   b. open when the relay coil is deenergized.
   c. open when the relay coil is energized.
   d. contacts that open a circuit.

8. Normally closed relay contacts are represented by the symbol: _____

   a. ⊥   c. ⊁⊬

   b. ⊥   d. ⊣⊢

9. A sealing circuit bypasses _____
   a. the armature circuit.
   b. the field circuit.
   c. the ON pushbutton contacts.
   d. the relay coil.

10. Elementary control diagrams are read from _____
    a. top to bottom.
    b. bottom to top.
    c. right to left.
    d. field to armature circuit.

THE DC COUNTER EMF MOTOR CONTROLLER AND DC VARIABLE SPEED MOTOR DRIVE

UNIT 12

OBJECTIVES

After studying this unit, the student will be able to

- explain the operation of the counter emf method of acceleration for a direct-current motor.
- make use of elementary wiring diagrams, panel wiring diagrams, and external wiring diagrams.
- explain the ratings of starting and running protection devices.
- describe the operating principles of dc variable speed motor drives.
- state how above and below dc motor speeds may be obtained.
- list the advantages of dc variable speed motor drives.
- describe how solid-state devices may replace rheostats.
- make simple drawings of dc motor drives.
- list the advantages of using thyristors.

Although manual starters are still used, most industrial applications use automatic motor control equipment to minimize the possibility of errors in human judgment. To install and maintain automatic motor control equipment the electrician must be familiar with three kinds of electrical circuit diagrams:

- elementary wiring diagram
- panel wiring diagram
- external wiring diagram

The elementary wiring diagram uses symbols and a simple plan of connections to illustrate the scheme of control and the sequence of operations.

The panel wiring diagram shows the electrical connections throughout all parts of the controller panel and indicates the external connections. All of the control elements are represented by symbols but are located in the same relative positions on the wiring diagram that they actually occupy on the control panel. Because of the maze of wires shown on the panel wiring diagram, it is difficult to use for troubleshooting or to obtain an understanding of the operation of the controller. For this reason, the elementary wiring diagram presents the sequence of operations of the controller and the panel diagram is used to locate problems and failures in the operation of the controller.

The external wiring diagram shows the wiring from the control panel to the motor and to the pushbutton stations. This diagram is most useful to the worker who installs the conduit and the wires between the starter panel and the control panel and motor.

COUNTER ELECTROMOTIVE FORCE METHOD OF MOTOR ACCELERATION CONTROL

The counter emf across the armature is low at the instant a motor starts. As the motor accelerates, this counter emf increases. The voltage across the motor armature can be used to activate relays which reduce the starting resistance when the proper motor speed is reached.

Starting and Running Protection for a Counter Emf Controller

Starting protection for a counter emf controller is provided by fuses in the motor feeder and branch-circuit line of the motor circuit. These fuses are rated according to *Article 430* of the NEC.

Running protection for a counter emf controller is provided by an overload thermal element connected in series with the armature. The thermal element is rated at 115 to 125 percent of the full-load armature current. As covered in NEC *Sections 430-32* and *430-34*, if the current exceeds the percent of the rated armature current value, heat produced in the thermal element causes the bimetallic strip to open or trip the thermal contacts which are connected in the control circuit. The value of current during the motor startup period does not last long enough to heat the thermal element sufficiently to cause it to open.

COUNTER EMF MOTOR CONTROL CIRCUIT

Starting the Motor (Refer to Figure 12-1)

Close the main line switch before pressing the start button. After the start button is pressed, control relay M becomes energized, thus closing contacts 2–3 of contactor M. The control circuit is complete from L_1 through the thermal overload (OL) con-

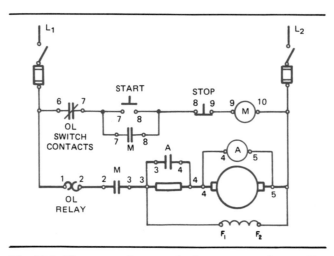

Fig. 12-1 Elementary diagram of a dc counter emf controller

tacts 6–7, through the start button contacts 7–8, through the normally closed stop button contacts 8–9 to L_2. The lower auxiliary sealing contacts 7–8 of relay M also close and bypass the start button. As a result, the start button may be released without disturbing the operation.

When the main contacts 2–3 of contactor M are closed, the motor armature circuit is complete from L_1 through overload thermal element contacts 1–2, through contacts 2–3 of relay M, through the starting resistor, and through armature leads 4–5 to L_2. The shunt field circuit $F_1 - F_2$ is connected in parallel with the armature circuit. Contacts 3–4 of counter emf contactor A remain open at startup because a high inrush current establishes a high voltage drop across resistor 3–4. This leaves only a small voltage drop across the armature and "A" contactor coil until acceleration is achieved.

Connecting the Motor across the Line

The counter emf generated in the armature is directly proportional to the speed of the motor. As the motor accelerates, the speed approaches the normal full speed and the counter emf increases to a maximum value. Relay A is calibrated to operate at approximately 80 percent of the rated voltage. When contacts 3–4 of relay A close, the starting resistance 3–4 is bypassed and the armature is connected across the line.

Running Overload Protection

When the load current of the armature exceeds the rated allowable percent of the full load current, the overload thermal element (contacts 1–2) opens contacts 6–7 in the control circuit. Control relay M is deenergized and main contacts 2–3 of M open and disconnect the motor from the line.

Stopping the Motor

When the stop button is pressed the control circuit is broken. The same shutdown sequence occurs as in the case of the overload condition discussed previously. The sealing circuit 7–8 is broken in each case.

The advantage of this type of automatic starter is that it does not supply full voltage across the armature until the speed of the motor is correct. The starter eliminates human error which may result from the use of a manual starter.

PANEL WIRING DIAGRAM

Figure 12-2 shows the same counter emf control circuit presented in figure 12-1. However, the panel wiring diagram locates the wiring on the panel in relationship to the actual location of the equipment terminals on the rear of the control panel. Troubleshooting or checking of original installations requires an accurate comparison of the elementary and panel diagrams. It is recommended that the electrician use a system of checking connections on the diagram with the actual panel connections. For example, a colored pencil may be used to make check marks on the diagram as each connection is properly traced on the panel and compared to the diagram.

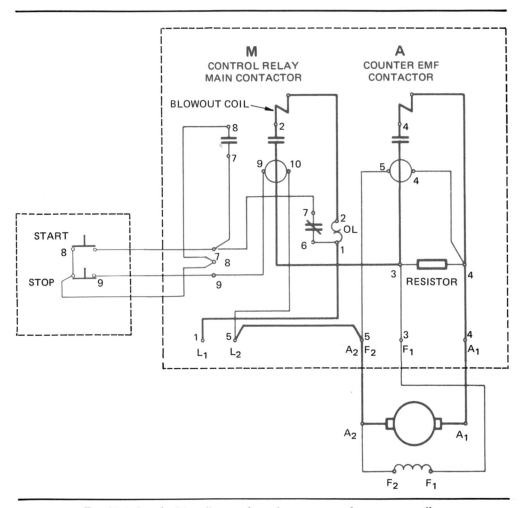

Fig. 12-2 Panel wiring diagram for a dc counter emf motor controller

CONDUIT OR EXTERNAL WIRING PLAN

All necessary external wiring between isolated panels and equipment is shown in the conduit plan (figure 12-3). The proper size of conduit, size and number of wires, and destination of each wire is indicated on this plan. An electrician refers to this plan when completing the actual installation of the counter emf controller.

DC ADJUSTABLE SPEED DRIVES

Dc adjustable speed drives are available in convenient units that include all necessary control and power circuits. These packaged drive units generally operate on ac for both rotating equipment and solid-state controlled equipment.

Some machinery requirements are so precise that the ac motor drive alone may not be suitable. In such cases, dc motors provide characteristics that are not available on ac motors. A dc motor with adjustable voltage control is very versatile and can be adapted to a large variety of applications.

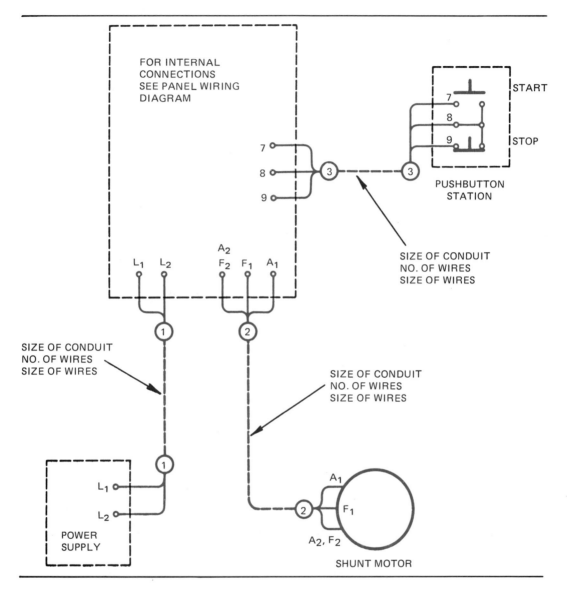

Fig. 12-3 Conduit or external wiring plan for a counter emf controller

In the larger horsepower range, the motor-generator set has been one of the most widely used methods of obtaining variable speed control. The set consists of an ac motor driving a dc generator to supply power to a dc motor. Such motor-generator set drives, called Ward-Leonard Systems, control the speed of the motor by adjusting the power supplied to the field of the generator, and, as a result, the output voltage to the motor (figure 12-4). The generator field current can be varied with rheostats, as shown, or by variable transformers supplying a dc rectifier, or automatically with the use of solid-state controls. When it is desirable to control the motor field as well, similar means are used.

The speed and torque of the system shown in figure 12-4 can be controlled by adjusting the voltage to the field, or to the armature, or both. Speeds *above* the motor

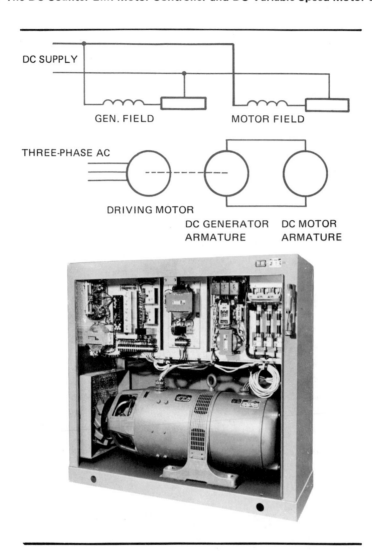

Fig. 12-4 A) Basic electrical theory of a dc motor-generator variable speed control system. B) Packaged motor-generator with dc variable speed control system supplied from ac *(Photo courtesy Square D Company)* **(From Alerich, *Electric Motor Control*, copyright 1983 by Delmar Publishers Inc.)**

base speed (nameplate speed) are obtained by weakening the motor shunt field. Speeds *below* the motor base speed are obtained by weakening the generator field. As a result, there is a decrease in the generator voltage supplying the dc motor armature. The motor should have a full shunt field for speeds lower than the base speed to give the effect of continuous control, rather than step control of the motor speed.

The motor used to furnish the driving power may be a three-phase induction motor, as shown in figure 12-4. After the driving motor is started, it runs continuously at a constant speed to drive the dc generator.

The armature of the generator is coupled electrically to the motor armature as shown. If the field strength of the generator is varied, the voltage from the dc generator can be controlled to send any amount of current to the dc motor. As a result, the motor can be made to turn at many different speeds. Because of the inductance of the dc fields and the time required by the generator to build up voltage, extremely smooth acceleration is obtained from zero r/min to speeds greater than the base speed.

The field of the dc generator can be reversed automatically, or manually, with a resulting reversal of the motor rotation.

The generator field resistance can be changed automatically by the use of SCRs (or thyristors) or time-delay relays operated by a counter EMF across the motor armature. The generator field resistance can also be changed manually.

Electrically controlled variable speed motor drives offer a wide choice of speed ranges, torque, and horsepower characteristics. They provide a means for controlling acceleration and deceleration, and methods of automatic or manual operation. A controlling tachometer feedback signal may be driven by the dc motor shaft. This is a system refinement to obtain a preset constant speed. This method depends upon the type of application, the speed, and the degree of response desired. In addition to speed, the controlling feedback signal may be set to respond to pressure, tension, shock, or some other transducer function.

One of the most advantageous characteristics of the motor-generator set drive is its inherent ability to regenerate. In other words, when a high inertia load overdrives the motor, the dc motor becomes a generator and delivers power back. For example, assume the dc motor is running at base speed. If the generator voltage is decreased by adjusting the rheostat, the motor counter voltage will be higher than the generator voltage and the current reverses. This action results in reverse torque in the machine and the motor slows down. This process is called dynamic braking. This dynamic feature is very desirable when used on hoists for lowering heavy loads, metal working machines, textile and paper processing machines, and for general industry for the controlled stopping of high inertia loads. Multiple motor drives are also accomplished with this type of motor-generator drive.

Motor-generator set drives using automatic regulators have been used for years for nearly every type of application. A higher degree of sophistication in controls has been developed, making it possible to meet almost any desired level of precision or response.

STATIC MOTOR CONTROL DRIVES

Despite the widespread acceptance and use of the motor-generator drives, rotating machines are required to convert ac to mechanical power. As a result, the combined efficiency of the set is rather low; it requires the usual rotating machine maintenance, and it is noisy. *Static* dc drives now being used have no moving parts in the power

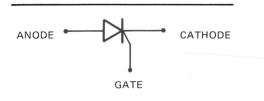

ANODE   CATHODE

GATE

Fig. 12-5 Symbol for a silicon-controlled rectifier (SCR),
or thyristor, the heart of the control package that converts
ac to dc to provide smooth control of motor output.

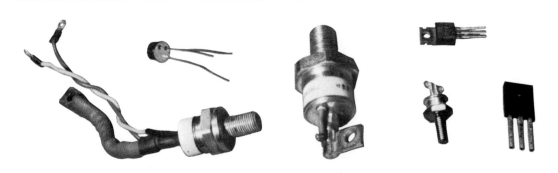

Fig. 12-6 SCRs in different sizes and configurations (From Herman/Alerich, *Industrial Motor Control,* copyright 1985 by Delmar Publishers Inc.)

conversion equipment that converts (rectifies) and controls the ac power (figures 12-5 and 12-6).   The solid-state devices are used for controlled conversion of ac line power to dc.

The basic theory for obtaining dc motor speeds below and above base speed are the same as with a motor-generator set. It is only the method of controlling the voltages and field strengths that differs. For example, in Figure 12-7, the armature is supplied with dc from an ac source. The ac is rectified by the use of the thyristor in the controlled circuit to obtain dc. The gate of the thyristor can be preset to trigger (turn on) the device at the selected power value (portion) of the half wave, thereby controlling the motor below base speed. Figure 12-7 is a simplified circuit for the purpose of illustration. The field strength would be held at its fullest strength in a similar manner. For above motor base speed, the field control can weaken field strength with full armature voltage. The feedback tachometer will maintain a preset speed.

In figure 12-7, the SCR is controlled by the setting of the potentiometer, *speed control.*   This varies the "on" time of the thyristor per ac cycle, and thus varies the

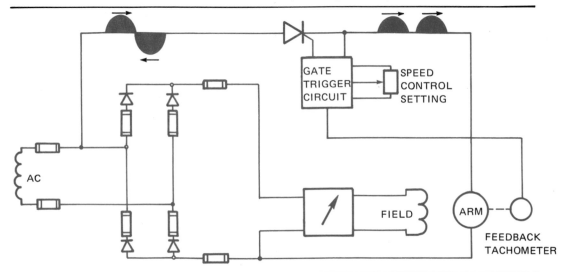

Fig. 12-7 Single-phase, half-wave armature controlling a small motor

amount of average current flow to the armature. When speed control above the base speed is required, the rectifier circuit in the field is controlled by rectifiers and SCRs, rather than diodes.

The SCR, or thyristor, can control all of the positive waveform or voltage through the use of a method called *phase shifting*. It is beyond the scope of this text to cover the theory of the method.

The SCR is probably the most popular solid-state device for controlling large and small electrical power loads. The SCR is a controlled rectifier which conducts, or does not conduct, an electric current. It will not conduct when the voltage across it is in the wrong direction. It will conduct only in the forward direction when the proper signal (voltage) is applied to the gate terminal. This may be a pushbutton operation.

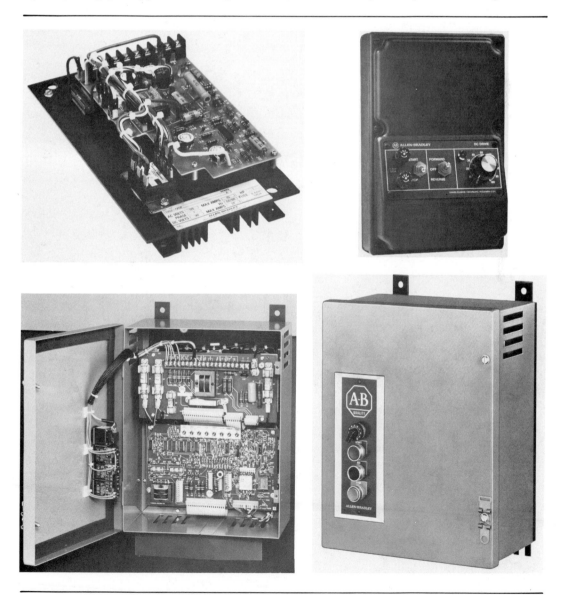

Fig. 12-8 Ac motor drive control *(Photos courtesy of Allen-Bradley Company)*

The gate will turn the SCR on but will not turn it off in a dc circuit. To turn the anode-cathode section of the SCR on (close the switch), the gate must be connected to the same polarity as the anode. Once the gate has turned the SCR on, it will remain on until the current flowing through the power circuit (anode-cathode section) is either interrupted or drops to a low enough level to permit the device to turn off. The *holding current,* or *maintaining current,* is the amount of current required to keep the SCR turned on. The SCR performs the same function as a rheostat in controlling motor field strengths or voltage to an armature. It is similar to a variable resistance, since it can be adjusted throughout its power range. The SCR has greater acceptance than a rheostat since it is smaller in size for the same current rating, is more energy efficient, and is cheaper.

Figure 12-8 illustrates two sophisticated, single-phase, packaged, static control ac motor drive controls. They are shown with their wiring exposed and in their boxes. More elaborate units are available for three-phase ac power supplies.

ACHIEVEMENT REVIEW

Select the correct answer for each of the following statements and place the corresponding letter in the space provided.

1. The least important plan or diagram in troubleshooting motor controls is probably the   _____
   a. elementary plan.
   b. panel diagram.
   c. external conduit plan.
   d. layout of the area in which the controllers are installed.

2. The best diagram to use to determine how a controller operates is the   _____
   a. elementary plan.   c. external plan.
   b. panel plan.   d. architectural plan.

3. The physical location of control wires is shown on the   _____
   a. elementary plan.
   b. architectural plan.
   c. conduit plan.
   d. panel wiring diagram.

4. The dc counter emf controller results from the automatic actions of the   _____
   a. applied voltage.
   b. changing voltage across the armature.
   c. changing voltage across the field.
   d. starting current.

5. Overload protection is the same as   _____
   a. starting protection.   c. electrical protection.
   b. mechanical protection.   d. running protection.

6. Overload contacts open the circuit when the motor current reaches _____
    a. 85 percent of full load.
    b. 100 percent of full load.
    c. 125 percent of full load.
    d. 150 percent of full load.

7. In the event a motor is allowed to exceed the permissible current value, it is protected by _____
    a. starting protection.
    b. fuses.
    c. an overload thermal element.
    d. the stop button.

8. With the disconnect switch closed, the shunt field in figure 12-1 is placed across the line when the _____
    a. A contact closes.
    b. disconnect switch is closed.
    c. M contact closes.
    d. start button closes.

9. In figure 12-1, contact A is closed when the _____
    a. start button is closed.
    b. stop button is opened.
    c. A coil is deenergized.
    d. A coil is energized.

10. The motor in figure 12-1 is placed across the line when _____
    a. the start button is closed.
    b. the disconnect switch is closed.
    c. contact A is closed.
    d. contact M is closed.

11. What is the dc motor base speed? _____

_____

12. How is the speed of a dc motor controlled *above* the base speed? _____

_____

_____

_____

13. How is the speed of a dc motor controlled *below* the base speed? _____

_____

_____

_____

14. How may an SCR replace a rheostat? _____

_____

_____

15. List the advantages of using thyristors in the motor drive control? _____

_____

_____

THE DC VOLTAGE DROP ACCELERATION CONTROLLER

OBJECTIVES

After studying this unit, the student will be able to

- state the purpose of a dc voltage drop acceleration controller.
- explain the principle of operation of lockout relays in a dc voltage drop acceleration controller.
- list, in sequence, the steps in the operation of an acceleration controller.
- connect an acceleration controller to a dc motor.

Large dc motors must be accelerated in controlled steps. A series of resistors or tapped resistors connected to lockout relays can be used to provide uniform motor acceleration. Since the starting current of a motor is high, the voltages across the series starting resistors are high and the voltage across the motor armature is low. As the motor accelerates and the counter emf increases, the armature current and the voltage drop across the series starting resistors decrease. Lockout relays connected across these resistors are calibrated to short out the series starting resistors as the motor speeds up.

CIRCUIT FOR THE VOLTAGE DROP ACCELERATION CONTROLLER

The Lockout Relay

For this circuit, motor acceleration is divided into three steps due to the actions of three lockout relays. A lockout relay has two coils, as shown in figure 13-1. For each relay shown in figure 13-2, coil A is connected across the line and serves as the operating coil of the relay. Coil LA of each relay is connected across one starting resistor. Since there are three starting resistors, three lockout relays are used to provide three steps of acceleration. The voltage drop across the resistors is high enough during the starting period to prevent the main control coil A of each relay from closing the A contacts which shunt or bypass the starting resistors. This action is called *lockout*.

Starting the Motor

When the start button in figure 13-2 is pressed, the control relay M is energized. The main contacts 9–10 close and complete the armature circuit through the thermal element and the three resistors. The shunt field is connected across the line. Contacts 3–6 and 6–4 seal in the start button contact. Contacts 3–6 also energize coil 1A. The large value of starting current through the starting resistance produces a large voltage drop across each section of the starting resistance. As a result, there is a large current to the lockout coils 1LA, 2LA, and 3LA and the accelerating contactors are held open.

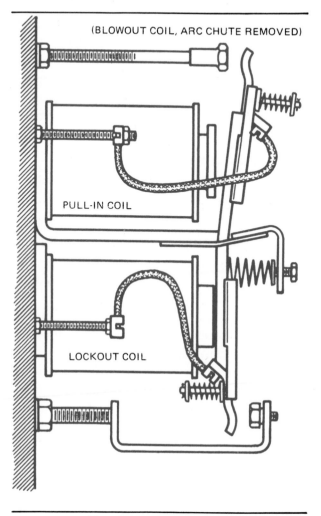

(BLOWOUT COIL, ARC CHUTE REMOVED)

PULL-IN COIL

LOCKOUT COIL

Fig. 13-1 A lockout relay

Contacts 4–6 are used to prevent coils 1A, 2A, and 3A from energizing when the start button is depressed. These coils receive power only when coil M is energized so they are not energized before coils 1LA, 2LA, and 3LA.

Acceleration

As the motor accelerates, the counter emf increases in the armature and the armature current decreases. With reduced current in the lockout coil of the relay, the pull of the lockout coil on the relay armature is less than the pull of the operating coil 1A. Therefore, relay 1A closes and contacts 1A bypass resistor R_1. Relay coil 2LA remains open due to the voltage drop across R_2, while relay coil 3LA remains open due to the voltage drop across R_2 and R_3. As R_1 is bypassed, the current again increases and the voltage drop across R_2 is high enough to keep the contacts of relay 2A open. However, as the motor continues to accelerate, the armature current decreases and lockout coil 2LA allows relay 2A to close contacts 2A and bypass resistor R_2.

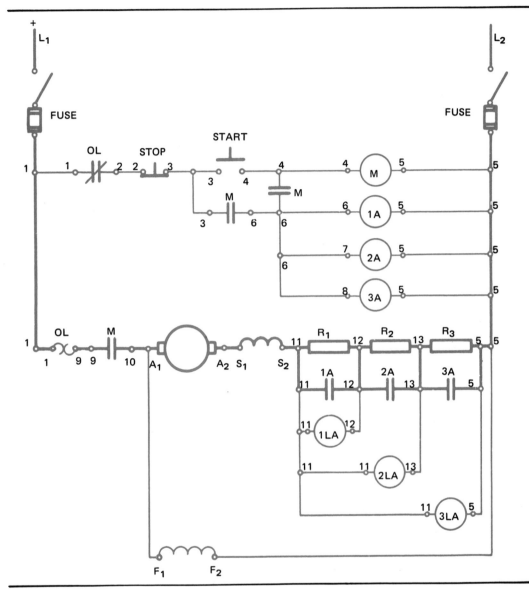

Fig. 13-2 Elementary diagram for the dc voltage drop acceleration controller

Connecting the Motor across the Line

Resistor R_3 is bypassed in the same manner as R_1 and R_2. The motor is accelerated to normal speed in three steps. It is connected directly across the line when resistor R_3 is cut out.

Running Overload Protection

A sustained overload greater than the permissible percentage of the full-load current causes the thermal element to open overload contacts 1–2 which, in turn, causes coil M to become deenergized. Contacts 9–10 are then opened and the motor is shut down. Manual shut-down protection is obtained by pressing the stop button.

ACHIEVEMENT REVIEW

Select the correct answer or answers for each of the following statements, and place the corresponding letter or letters in the space provided.

1. High motor starting current will
   a. increase the field current.
   b. decrease the speed.
   c. increase the voltage drop across the starting resistors.
   d. decrease the voltage drop across the starting resistors.

2. The voltage drop acceleration method of speed control is particularly suited to
   a. braking of motors.
   b. starting small motors.
   c. starting large motors in one step.
   d. starting large motors in three steps.

3. Large dc motors are started with
   a. single-step acceleration.
   b. multiple-step acceleration.
   c. elementary counter emf controllers.
   d. auxiliary motors.

4. The lockout relay has
   a. one main coil.
   b. three coils.
   c. one closing and one lockout coil.
   d. two coils wound in series.

5. The main coil in a lockout relay is connected across the source and the lockout coil is energized by the
   a. counter emf.
   b. voltage drop across the starting resistor.
   c. total current in armature.
   d. current in the field circuit.

6. The lockout coil of the lockout relay weakens and causes the relay contacts to close when the
   a. motor speed drops.
   b. counter emf rises and reduces the motor current.
   c. counter emf decreases.
   d. starting resistance is cut out.

7. A decreasing starting current causes the lockout relay to
   a. cut in the starting resistance.
   b. cut out the starting resistance.
   c. increase the field resistance.
   d. decrease the field resistance.

8. Motor circuit protection during the starting period is provided by the

   a. thermal unit.   c. distribution line fuses.

   b. branch fuses.   d. motor circuit breaker.

_____

9. A running load in excess of 150 percent of the full-load current causes the

   a. main line circuit to open before the control circuit.

   b. control circuit to be opened.

   c. speed to decrease.

   d. field rheostat to be bypassed.

_____

10. The panel wiring is completed by referring to the

   a. relay symbols.   c. conduit plan.

   b. elementary diagram.   d. motor schematic diagram.

_____

STUDENT ACTIVITY

Complete the conduit plan (figure 13-3) using the panel wiring diagram (figure 13-4) as a reference.

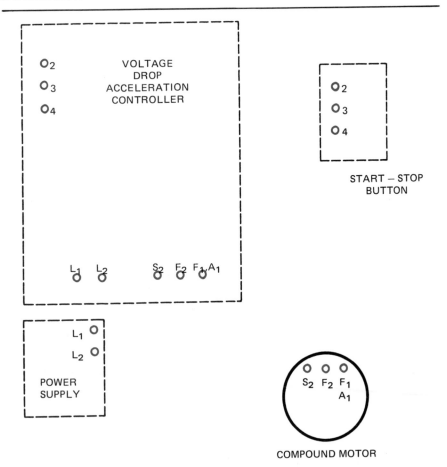

Fig. 13-3 Conduit plan, external wiring for a dc voltage drop acceleration controller

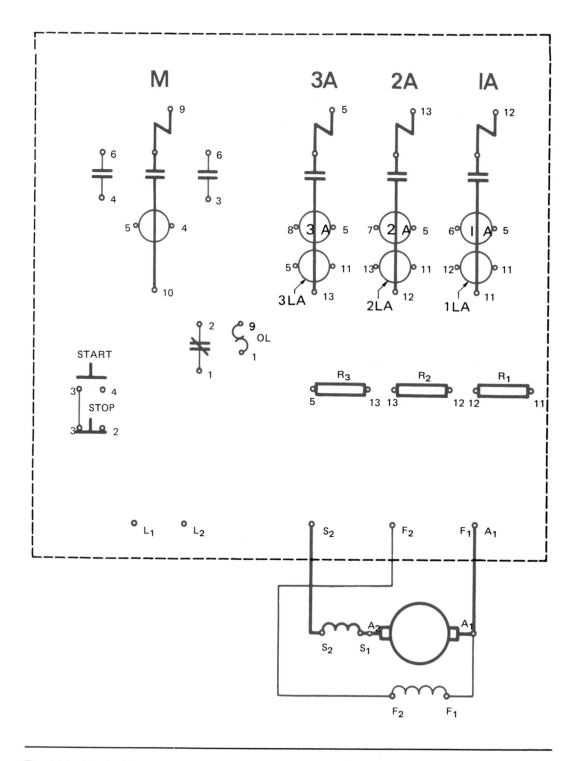

Fig. 13-4 Panel wiring diagram for the dc voltage drop acceleration controller with pushbuttons in the hinged cover

THE DC SERIES LOCKOUT RELAY ACCELERATION CONTROLLER

OBJECTIVES

After studying this unit, the student will be able to

- describe the operation of a dc series lockout relay.
- list the steps in the operating sequence of a series lockout relay acceleration controller.

A series lockout relay with its two coils is shown in figure 14-1. The lockout coil prevents the relay contacts from closing during the period of high motor starting current. The pull-in coil closes the relay contacts when the motor has accelerated and the starting current is reduced.

Note: In another type of relay, the magnetic gap is changed by shims located in back of the coil, rather than by an adjusting screw.

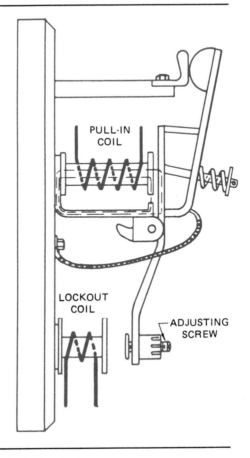

Fig. 14-1 Series lockout relay

OPERATION OF THE SERIES LOCKOUT ACCELERATION CONTROLLER

Figure 14-2 is the elementary diagram of a controller which uses series lockout relays to produce the necessary time delay or step control in motor acceleration.

Coils 1A, 2A, and 3A are the pull-in coils of the three relays, while coils 1HA, 2HA, and 3HA are the lockout coils of the relays.

When the start button is pressed, coil M is energized, and the main contactor, M, (1) seals the bypass auxiliary contact around the start button, and (2) allows current to flow in the following sequence: from L_1 through contact M, the motor armature, the series field, coil 1A, coil 1HA, resistors R_1, R_2, and R_3, and coil 3HA to L_2.

The large starting current through coil 1HA produces a large magnetic effect in this coil. Since this effect is larger than that of coil 1A, the relay is held open due to the fact that the magnetic path of the pull-in coil has a small amount of iron. As a result, this coil becomes saturated at high values of current. The magnetic circuit of the lockout coil has a larger amount of iron and, therefore, does not tend to become saturated at a high current value.

As the motor accelerates and the starting current decreases, the pull of the lockout coil becomes less than the pull of the pull-in coil and contactor 1A closes. Resistor R_1 and lockout coil 1HA are bypassed and thus increased current is allowed through lockout relay coils 2A and 2HA. A cycle of operation similar to that which occurred for relay 1A

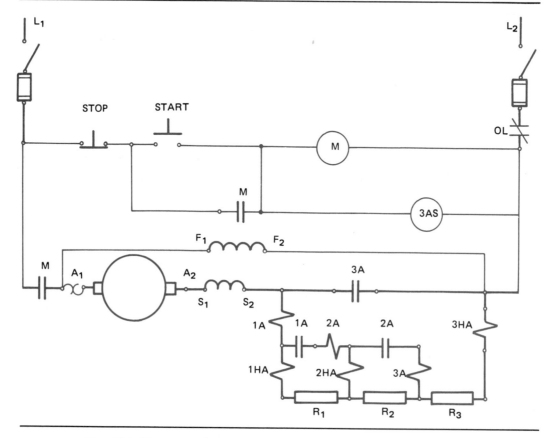

Fig. 14-2 Elementary diagram for the series lockout acceleration controller

now takes place for relay 2A. The current path is now from L_1 through M, the armature and series field, coil 1A, contact 1A, coils 2A and 2HA, resistors R_2 and R_3, and coil 3HA to L_2. The large current through lockout coil 2HA produces more magnetic pull than is present at coil 2A with the result that contact 2A is held open.

Finally, as the motor accelerates, all resistors and relays are shunted out. This should immediately cause all relay contacts to drop out, except for the fact that coil 3AS is an auxiliary shunt coil which acts on contactor 3A. Coil 3AS is strong enough to hold contact 3A closed after it has been pulled into contact, but is not strong enough to cause 3A to close its contacts without the aid of coil 3A.

When there is a heavier load on the motor, acceleration takes place over a longer period of time.

ACHIEVEMENT REVIEW

1. How is control relay M in the series lockout acceleration controller held in when the start button is released?

    _____

    _____

2. How does the lockout relay short out a starting resistor when the motor is accelerated?

    _____

    _____

3. What will happen if a break occurs in the coil of 3A?

    _____

    _____

    _____

4. A motor accelerates through two steps only. The speed is below normal and the motor voltage is low at the armature terminals. What is the probable cause of the problem?

    _____

    _____

    _____

5. What is the purpose of the 3AS coil of the lockout relay?

    _____

    _____

    _____

DYNAMIC BRAKING WITH A DC MOTOR REVERSAL CONTROL

OBJECTIVES

After studying this unit, the student will be able to

- list the steps in the operation of a dc motor control with interlocked forward and reverse pushbuttons.
- explain the principle of dynamic braking.
- describe the operation of a counter emf motor controller with dynamic braking.

Industrial motor installations often require that motors be stopped quickly and that the direction of rotation be reversed immediately after stopping. To achieve this operation, electrically and mechanically interlocked pushbutton stations connected to relays are used to disconnect the armature from the supply source. The armature is then connected to a low value of resistance. Because the inertia of the armature and connected load causes the armature to continue to revolve, it acts as a loaded generator. As a result, the armature is slowed in speed. This action is called *dynamic braking.*

Reversal of motors and dynamic braking are operations used in special equipment such as cranes, hoists, railway cars, and elevators.

MOTOR REVERSAL CONTROL

A motor is reversed by reversing the armature connections. The type of compounding is not affected by this method of obtaining reversal.

The pushbutton control station illustrated in figure 15-1 is the type used for motor reversal. The forward and reverse buttons are mechanically interlocked so that it is not possible to operate these buttons at the same time.

Description of Operation

Forward Starting. When the forward button is pressed, the normally open forward contacts close and the normally closed forward contacts open.

Fig. 15-1 Pushbutton control station
(Courtesy General Electric Co.)

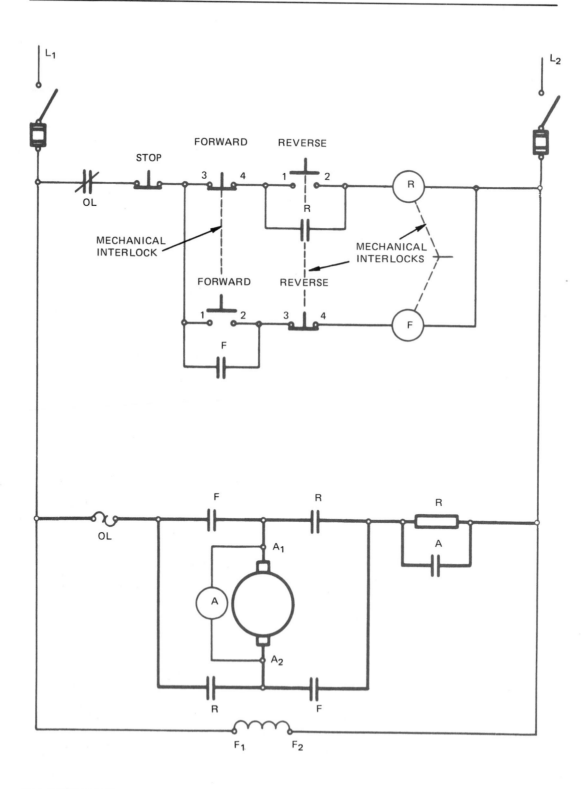

Fig. 15-2 Electrically and mechanically interlocked control and power circuit for reversing motor

The control circuit is shown in figure 15-2. The forward contactor coil is energized from L_1 through the overload contacts, stop button, forward pushbutton contacts 1–2 (when closed), and reverse button contacts 3–4 through the forward contactor coil to L_2. The forward contacts F seal in the forward pushbutton. In the power circuit (figure 15-2) the F contacts of the forward energized contactor close, and thus complete the armature circuit through the starting resistance. The normal counter emf starter sequence of operations then continues to completion.

Reverse Operation. If the reverse pushbutton is pressed, contacts 3–4 of the reverse button open, and thus deenergize the forward contactor coil F. In addition, the F contacts are opened as well as the sealing contacts F. Pressing the reverse button also completes the circuit of the reverse contactor coil R which closes the R contacts. The motor armature circuit is now complete from L_1 to A_2 and A_1 to L_2 (figure 15-2). The armature connections are reversed and the armature rotates in the opposite direction. It is impossible for the reverse contacts to close until the forward contacts are open, due to the electrical and mechanical interlocking system used in this type of control circuit. The mechanical interlocks are shown by the broken lines between the R and F coils in figure 15-2.

Dynamic Braking

The purpose of dynamic braking is to bring a motor to a quicker stop. To do this, there must be a method to use quickly the mechanical energy stored in the motion of the armature after the main switch is opened. One method is to change the function of the motor to that of a generator. (A generator converts mechanical energy into electrical energy.) At the instant the motor is disconnected from the line, a resistor is connected across the motor armature. The resistor loads the motor as a generator, dissipates the mechanical energy, and slows the motor quickly.

DYNAMIC BRAKING USED IN A COUNTER EMF CONTROLLER

As an example, the principle of dynamic braking is shown by following the steps in the operation of an elementary counter emf controller (figure 15-3).

This analysis emphasizes the dynamic braking operation rather than the details of the circuit which were presented previously. The dynamic braking coil (DBM) is designed so that its only function is to insure a positive closing of the normally closed dynamic braking contacts 9–10. If the main coil M is energized, the dynamic braking contacts 9–10 open and contacts 8–9 of M close, although the dynamic braking coil is also energized. The dynamic braking coil is a little weaker than the M coil.

When the start button is pressed, the control coil M is energized, contacts 8–9 of M close, and the motor starts and accelerates up to normal speed by the counter emf method. At the instant the M control relay is energized, the main, normally closed, dynamic braking contacts 9–10 open. As a result, the dynamic brake resistor connection across the armature is broken.

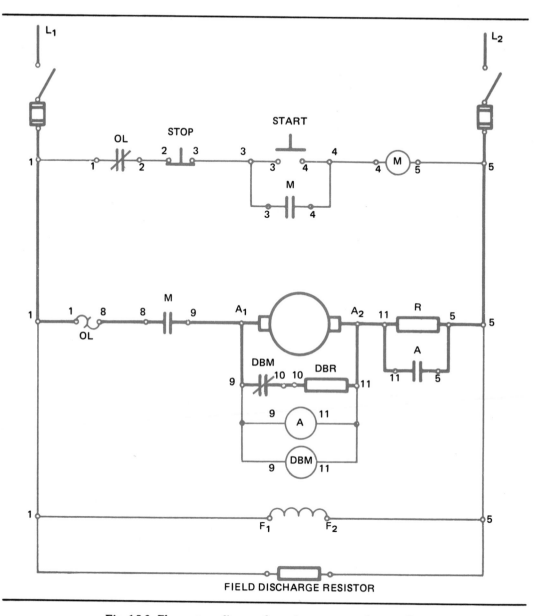

Fig. 15-3 Elementary diagram for a dc counter emf controller

Field Discharge Resistor

For most motor control circuits, the pressing of the stop button disconnects the entire motor circuit from the line and the field discharges through the armature. For the counter emf controller with dynamic braking, the armature is disconnected from the field when the stop button is pressed. A discharge resistor connected across the field replaces the armature as the dissipator of field energy when the main switch is opened.

Stopping

When the stop button is pressed, relay control coil M is deenergized, and M contacts 8–9 open the armature circuit and close the dynamic braking contacts 9–10 .

These contacts connect the dynamic brake resistor directly across the armature. Since the shunt field is still connected across the line and receiving full excitation, the high counter emf generated in the armature causes a high load current through the dynamic brake resistor. The heavy load current dissipates the stored mechanical energy in the armature with the result that the motor slows to a stop. The braking action decreases as the armature speed decreases.

ACHIEVEMENT REVIEW

1. How can dc motors be reversed without changing the type of compounding?

   _____

   _____

2. What interlocking is necessary on the forward and reverse pushbuttons to avoid short circuits? _____

   _____

3. What will happen if the forward and reverse relays are energized at the same time?

   _____

   _____

4. How many contacts are required on the forward relay? _____

5. What are two applications requiring motor reversal? _____

   _____

   _____

6. How can a motor be shut down quickly without using a mechanical brake?

   _____

   _____

7. When is dynamic braking applied? _____

   _____

   _____

8. How does dynamic braking slow a motor? _____

   _____

   _____

9. If an electromagnetically operated brake is connected in series with the armature, how is it operated? _____

_____

_____

_____

_____

_____

10. What are two installations where dynamic braking is used? _____

_____

_____

_____

_____

SUMMARY REVIEW OF UNITS 9-15

OBJECTIVE

- To give the student an opportunity to evaluate the knowledge and understanding acquired in the study of the previous seven units.

A. Complete the following statements.

1. A dc motor starter is designed to limit the _____.

2. A three-terminal starting rheostat provides _____ protection for a shunt motor.

3. Since the holding coil of a four-terminal rheostat is connected across the source, this type of rheostat provides _____ protection to a motor.

4. A four-terminal rheostat may be used with a motor for which a wide range of _____ control is required.

5. Series motor starters provide either _____ protection or _____ protection.

6. A drum controller is used when the operator has _____ control of the motor.

B. Select the correct answer for each of the following statements and place the corresponding letter in the space provided.

7. The three-terminal starting box can be used with all but the   _____
   a. compound motor.
   b. shunt motor.
   c. series motor.

8. Interpretation of automatic circuits requires the recognition of   _____
   a. color.   c. parallel circuits.
   b. rating   d. electrical circuit symbols.

9. A piece of apparatus which contains a minimum of one set of contacts operated by a coil is called a   _____
   a. motor.   c. relay.
   b. magnet.   d. dynamo.

10. Normally closed contacts are
    a. open at all times.
    b. open when the relay coil is deenergized.
    c. open when the relay coil is energized.
    d. closed when the relay coil is energized.

    _____

11. An elementary automatic motor controller circuit diagram shows the
    a. actual wiring layout.
    b. schematic motor diagram.
    c. actual sequence of operations of the entire circuit.
    d. sequence of operations of the starting circuit only.

    _____

12. One type of automatic controller operates on the basis that the counter emf generated in a motor
    a. increases the starting current.
    b. energizes a starting relay.
    c. deenergizes a starting relay.
    d. reduces the field current.

    _____

13. Another type of controller accelerates the motor in steps by
    a. using a series of thermal elements.
    b. allowing starting resistor voltage drops to energize relays.
    c. operation of relays controlled by speed.
    d. cam controlled relays.

    _____

14. A controller is called a series lockout relay starter because the
    a. starting resistors are bypassed and locked out.
    b. relays are all connected in series.
    c. relay coils are connected in series with the armature and cause each relay to lock out (open).
    d. relay coils lock themselves out of the circuit.

    _____

15. A remote-controlled motor reversal requires
    a. one relay with two contacts.
    b. two single contactor relays.
    c. two double contactor relays.
    d. two interlocked relays.

    _____

16. The principle of dynamic braking involves
    a. the use of a dynamo as a brake.
    b. the use of an electromagnetic brake connected across the motor armature.
    c. connecting a low resistance across the motor armature.
    d. connecting a brake relay across the armature.

    _____

C.   Complete each of the statements at the left by selecting the letter of the appropriate phrase or phrases from the list at the right. Write the letter(s) in the space provided.

17.   Undervoltage protection is provided in   _____ a. quick stopping.
18.   Dynamic braking is used for   _____ b. large motors.
19.   Multiple-step acceleration is used for   _____ c. a three-terminal starting
       starting   rheostat.
20.   Automatic starters provide more uni-   _____ d. a four-terminal starting
       form acceleration than   rheostat.
21.   A sealing circuit is used in   _____ e. small motors.
                                                      _____ f. series motors.
                                                      _____ g. manual starters.
                                                      _____ h. control circuits.

22.   Match each of the following symbols with its description at the right. Place the letter of the symbol in the space provided.

a.   ┤├   1.   _____ Normally closed pushbutton

b.   ⊥   2.   _____ Thermal motor overload relay
      o   o

c.   ─Ⓜ─   3.   _____ Shunt relay coil

d.   o⊥o   4.   _____ Normally open contact

e.   ╫   5.   _____ Fuse

f.   ⌒χ⌒   6.   _____ Normally closed contact

g.   ─[▭]─   7.   _____ Normally open pushbutton

INTRODUCTION TO POLYPHASE CIRCUITS

OBJECTIVES

After studying this unit, the student will be able to

- define what is meant by polyphase systems.
- state the advantages in the generation and transmission of three-phase power.
- measure and calculate power in three-phase systems.
- calculate the power factor in three-phase systems.

Almost all power transmission uses the three-phase system. In the three-phase system, electrical energy originates from an alternator which has three main windings placed 120 degrees apart. A minimum of three wires is used to transmit the energy generated.

A polyphase system, therefore, is the proper combination of two or more single-phase systems. In their order of usage, the most common types of polyphase systems are:

- three phase (used for power transmission)
- six phase (used for power rectification)
- two phase (used for power rectification)

Fig. 17-1 A transmission line substation *(Courtesy of General Electric Co.)*

Fig. 17-2 Steam turbine generators

Figure 17-1 shows equipment assembled in a transmission line substation, and figure 17-2 shows three-phase generators.

ADVANTAGES OF THREE-PHASE SYSTEMS

The advantages of three-phase systems apply to both the generation and transmission of electrical energy.

Generation

A three-phase generator may be compared to a gasoline engine. An eight-cylinder engine develops eight small pulses of power per cycle as compared to one large surge of power per cycle for a one-cylinder engine. Similarly, a three-phase generator generates energy in three windings per turn, rather than in just the one winding of a single-phase generator. In addition, the generator actually is smaller in physical dimensions than a single-phase generator of the same rating. Three-phase generation produces energy more smoothly than single-phase generation and provides for more economical use of space within the frame of the machine.

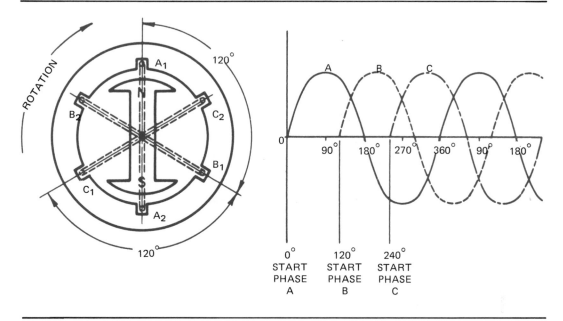

Fig. 17-3 Generation of three-phase electrical energy

Transmission Using Transformers

Three-phase transmission saves material, installation time, and maintenance costs. A three-phase, four-wire system can provide three 120-V lighting circuit lines, three 208-V single-phase circuits, and one 208-V, three-phase power line over four wires.

GENERATION OF THREE-PHASE ELECTRICAL ENERGY

Figure 17-3 shows the arrangement of the windings in a simple ac generator. The coils are spaced 120 electrical degrees apart. The voltage diagram shows the relationship of the instantaneous voltages as the rotating field poles turn in the direction indicated.

Three-Phase Winding Connections

The internal winding connections shown in figure 17-4 for a three-phase generator are arranged so that any of three or four wires may be brought out. That is, three-phase windings may be connected either in the star (wye) pattern (figure 17-5), or the delta pattern (figure 17-6).

Fig. 17-4 Schematic diagram of three-phase windings

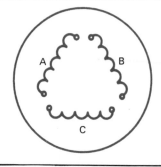

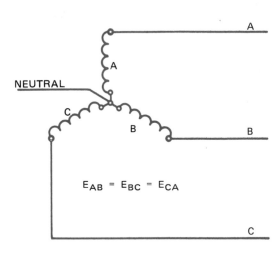

Fig. 17-5 Wye three-phase connection

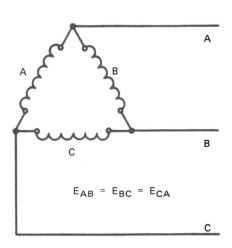

Fig. 17-6 Delta three-phase connection

SIX-PHASE CONNECTIONS

Six-phase power is usually applied to power rectifiers. The ac six-phase supply is converted from a three-phase power line by a bank of three transformers connected for six-phase on the secondary side of the line transformer. The double wye, six-phase connection shown in figure 17-7 is one method of obtaining six-phase power. The lines are brought out at the outer ends of the windings. The points (1-2-3-4-5-6) are displaced 60 electrical degrees apart from one another.

TWO-PHASE CONNECTIONS

In a two-phase connection (figure 17-8) the windings are spaced 90 degrees apart. Lines A_2 and B_1 are often connected to form a three-wire, two phase system. The A and B phase voltages are designed to be equal.

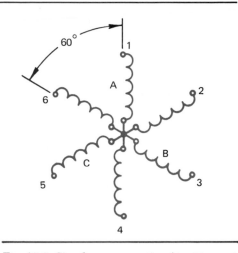

Fig. 17-7 Six-phase connection (double wye)

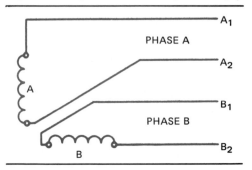

Fig. 17-8 Two-phase connection

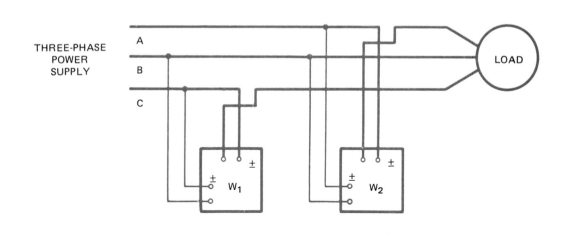

Fig. 17-9 Measurement of three-phase power using the two-wattmeter method

MEASUREMENT OF THREE-PHASE POWER

Three-phase power may be measured using either the two-wattmeter method or a polyphase wattmeter.

Two-Wattmeter Method

The reversing switches of both wattmeters must be set in the same direction. If one meter shows a negative reading, reverse the corresponding switch so that a positive reading is seen on the scale. If the switches are set in opposite directions, the lower reading is assumed to be negative.

If the power factor (PF) is 100%, $P_1 = P_2$, then $P_T = P_1 + P_2$.

If the PF is greater than 50% and less than 100%, then P_1 and P_2 are unequal. $P_T = P_1 + P_2$. (P_T = Total power.)

If the PF is 50%, one meter reads zero and $P_T = P - 0$.

If the PF is less than 50%, one meter has a negative reading and $P_T = P_1 - P_2$.

Note: The two-wattmeter method cannot be used in an unbalanced three-phase, four-wire system.

Polyphase Wattmeter Method

The connections of a polyphase wattmeter used to measure three-phase power are shown in figure 17-10. In the polyphase wattmeter, the torque produced by two current coils and two voltage coils causes the pointer to deflect and indicate total watts in the circuit, in one instrument.

CALCULATION OF THREE-PHASE POWER AND THE POWER FACTOR

Power Factor

The power factor is the ratio of true power to apparent power. A power factor of 100% is the best electrical system.

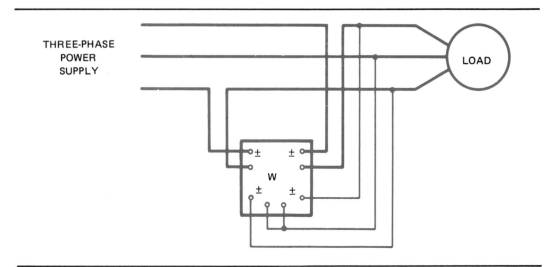

Fig. 17-10 Polyphase wattmeter connections for measurement of power in a three-phase circuit

$$PF = \frac{\text{True power}}{\text{Apparent Power}} = \frac{P}{\sqrt{3} \times E_L \times I_L} \qquad \text{Note: } \sqrt{3} = 1.73$$

This formula can be used only for *balanced three-phase circuits.*

A power factor meter can be used to measure the power factor in a three-phase circuit for both balanced and unbalanced conditions.

Power, then, is calculated using the following expression:

$$P = \sqrt{3} \times E_{\text{Line}} \times I_{\text{Line}} \times PF$$

The substitution of values in the foregoing formula determines the power in a three-phase *balanced* circuit only.

Three-phase equipment is designed to operate as a balanced load. A three-phase circuit containing a combination of single- and three-phase loads is very seldom balanced. A polyphase wattmeter must be used for the measurement of power in an unbalanced circuit.

ACHIEVEMENT REVIEW

For each of the numbered items, select the letter of the phrase from the following list which will complete the statement. Write the letter in the space provided.

a. two wattmeters

b. $P = \sqrt{3}\, E \times I \times PF$

c. polyphase wattmeter

d. three wires

e. four wires

f. five wires

g. six wires

h. more economical

i. rectification

j. two-phase power

k. three-phase power

l. more than one phase

m. 180 degrees

n. lighting

o. $E \times I \times PF$

1. Polyphase means   _____

2. Three generator windings spaced 120 degrees apart generate   _____

3. Two generator windings spaced 90 degrees apart generate   _____

4. Three-phase power is transmitted over a minimum of   _____

5. The six-phase connection is used for   _____

6. Three-phase power may always be measured by using a   _____

7. The formula for calculation of three-phase power is   _____

8. Unbalanced three-phase power can be measured with a   _____

9. The three-phase system is better than the single-phase line because it is   _____

10. Lighting circuits usually are connected to a three-phase line having   _____

THE THREE-PHASE WYE CONNECTIONS

OBJECTIVES

After studying this unit, the student will be able to

- diagram the proper connections for a wye-connection generator and transformers.
- state the applications of the wye-connection generator and transformers in three-phase distribution systems.
- compute the voltage and current values in various parts of the wye-connection circuit.

The star or wye connection is particularly suited for the distribution of power and lighting where one three-phase transmission line supplies the energy. All three transformers in the wye bank share the single-phase load as well as the three-phase load. The wye system also provides a grounded neutral with equal voltage between each phase wire and the neutral.

Two sets of voltages are available from the four-wire wye system: 120/208 and 277/480. The 120/208-V wye system is most commonly used for small industrial plants, office buildings, stores, and schools. In these applications, the main electrical need is for 120-V lighting and equipment circuits, and only a moderate amount of 208-V three-phase power load. The 277/480-V wye system is mainly used for large commercial buildings and industrial plants where there is a higher demand for power at 480 V, three phase, and lighting at 277 V, single phase.

The types of three-phase systems are named for the shape of the transformer secondary winding connections. The wye or star system, as shown in figure 18-1, is shaped as the letter Y. The wye or star connections are made by tying together the ends of the three transformer windings, labeled X_2, and bringing this termination out as the *neutral* wire. The remaining three unidentified conductors of the four-wire, three-phase system, labeled on the figure as A, B, C, are tied to the three X_1 ends, respectively. Alternators are connected in the same manner as shown in figure 18-1.

VOLTAGE RELATIONS

The voltage reading across any pair of line wires of a balanced three-phase wye connection is equal to the vector sum of the two-phase windings connected in series across the pair of lines.

For example, if the phase winding voltage is 120 V, the line voltage is 208 V.

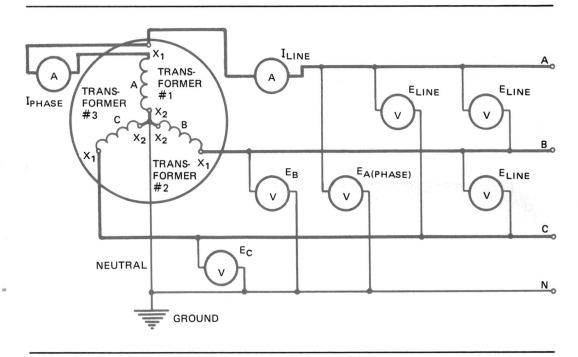

Fig. 18-1 Wye connections in an alternator, or three-phase transformer bank

$$E_{Line} = 1.73 \times 120 = 208 \text{ V}$$

The voltage from any line to a grounded neutral is the phase winding voltage and is usually called the *phase voltage*. Phase is represented by the Greek letter phi (ϕ).

CURRENT RELATIONS

The line current is the same as the phase current in a wye connection due to the fact that each phase winding is connected in series with its corresponding line wire (figure 18-1). Remember that current in a series circuit is the same throughout all parts of the circuit.

$$I_{Line} \ (I_L) = I_{Phase} \ (I_\phi)$$

APPLICATION

A four-wire transmission line usually originates from a transformer bank or a generator connected in wye. Both lighting and power circuits are connected to the four-wire system. Four circuits are served by four wires.

Lines A-B-C	Power
Lines A-Neutral	Three lighting circuits
B-Neutral	
C-Neutral	

In summary, the following statements are true of the three-phase wye connection:

$$E_{Line} = 1.73 \times E_{Phase}$$

$$I_{Line} = I_{Phase}$$

Lines A-B-C supply power circuits.

Lines to neutral serve lighting circuits.

Either three- or four-wire transmission lines are attached to a wye-connected generator or transformer bank.

WYE CONNECTION OF TRANSFORMER WINDINGS

A bank of three transformers can be connected in wye, delta, or other three-, six-, twelve-, or eighteen-phase arrangements. Figure 18-2 shows a conventional method of connecting three transformers in a three-phase wye arrangement. Compare this with figure 18-1.

Fig. 18-2 Transformer bank with a wye connection

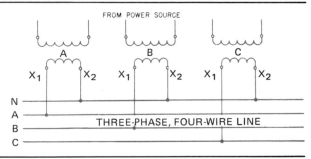

Unit substation consisting of two 4,500-kVA, 11,500- to 2,300- to 460-V, three-phase, 60-Hz transformers.

Fig. 18-3 An outdoor unit substation *(Courtesy of General Electric Co.)*

Fig. 18-4 Three-phase, high-voltage transformer bank

ACHIEVEMENT REVIEW

1. Indicate the number of the leads which must be connected to make a wye connection.

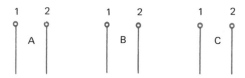

2. Where will the neutral be brought out in the diagram shown in question 1? Describe or use a sketch.

3. What two types of circuits are supplied by the three-phase, four-wire system?

   _____

   _____

   _____

4. Make a complete wye connection in the following diagram.

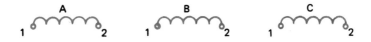

   N _____

   A _____

   B _____

   C _____

5. The phase current and phase voltage of each winding of an ac generator are 10 amperes and 100 volts, respectively. Determine the line voltage and current.

   _____   _____

   The following circuit is incomplete. Questions 6 through 10 are based on this circuit.

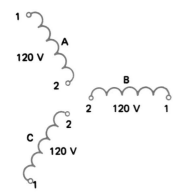

   _____ A
   _____ B
   _____ C
   _____ N

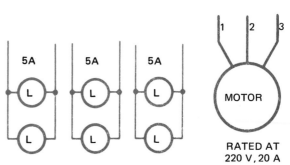

120-V LIGHTING CIRCUITS

RATED AT
220 V, 20 A

6. Complete the connections for a four-wire wye system.

7. Determine the line voltage.

8. Connect the motor for three-phase operation.

9. Connect the lamp banks for a balanced three-phase load.

10. Determine the phase current when only the lighting load is on the line.

THE THREE-PHASE DELTA CONNECTION

OBJECTIVES

After studying this unit, the student will be able to

- diagram the proper way to make a delta connection.
- state the applications of a delta-connected circuit in three-phase distribution systems.
- compute the voltage and current values in various parts of the delta-connection circuit.
- make a delta connection.

The delta connection, like the wye connection, is used to connect alternators, motors, and transformers. Delta is the Greek letter D which is shaped like a triangle (Δ). The delta connection takes its name from this symbol because of its triangular appearance. The schematic diagram of the winding connection of an alternator or secondary transformer bank shows the windings which are spaced 120 electrical degrees apart (figure 19-1).

CONNECTIONS

To make a delta connection, connect the beginning of one phase to the end of the next phase until the last and final connection is to be closed. DO NOT COMPLETE

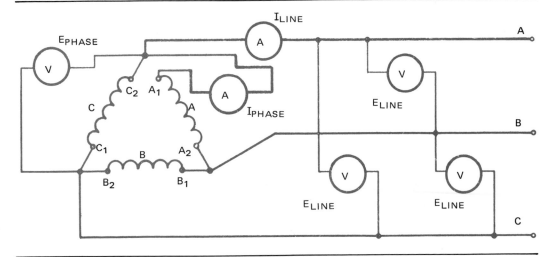

Fig. 19-1 Delta connection in an alternator (or three single-phase transformers)

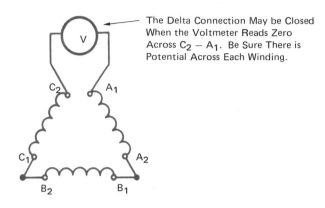

The Delta Connection May be Closed When the Voltmeter Reads Zero Across $C_2 - A_1$. Be Sure There is Potential Across Each Winding.

Fig. 19-2 Test for completion of the delta connection

THE DELTA CONNECTION until the voltage is measured across the last two ends (see $C_2 - A_1$ in figure 19-2).

Test for Completion of the Delta Connection

If the voltmeter reads zero across $C_2 - A_1$, the circuit may be closed (figure 19-2). If the voltmeter reads twice the voltage of the phase winding, reverse any phase and re-test. If a potential remains across $C_2 - A_1$, reverse a second phase and make a final voltage test before completing the delta connection. The phase windings must have potentials 120 electrical degrees apart.

VOLTAGE RELATIONS

The voltage measured across any pair of line wires of a balanced three-phase delta connection is equal to the voltage measured across the phase winding (see figure 19-1).

$$E_{Line} = E_{Phase}$$

CURRENT RELATIONS

Trace any line, such as line A, back to the connection point of phases C and A in a closed delta. The current in line A is supplied by phases A and C at the point of connection in the ac generator. Phases A and C are out of phase by 120 degrees. The line current, therefore, is the vector addition of the two phase currents. In a balanced circuit, the phase currents are equal. The line current is determined by the following formula.

$$I_{Line} = \sqrt{3} \times I_{Phase} \text{ or } 1.73 \times I_{Phase}$$

For example, if the phase current in each winding of a generator or transformer is 10 amperes, the line current is equal to $1.73 \times 10 = 17.3$ amperes.

$$I_{Line} = 1.73 \times 10 = 17.3 \text{ amperes}$$

APPLICATION

The delta connection may be used as the source of a three-wire transmission line or distribution system. The three-wire delta system is used when three-phase power on three conductors is required.

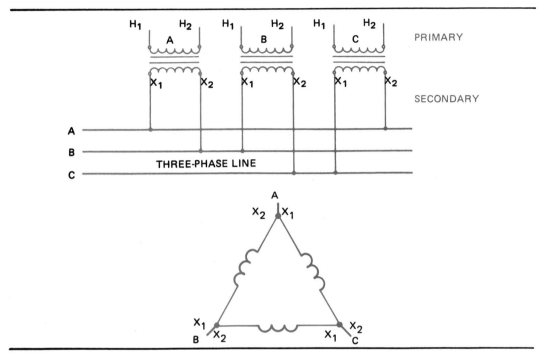

Fig. 19-3 Delta connections for a transformer bank

In summary, the following statements are true of the three-phase delta connection:

$$E_{Line} = E_{Phase}$$
$$I_{Line} = 1.73 \times I_{Phase}$$

DELTA CONNECTION OF TRANSFORMERS

Figure 19-3 shows the delta secondary connection of a bank of transformers.

ACHIEVEMENT REVIEW

1. As shown in the following diagram, six leads are brought out of a three-phase alternator and marked as indicated. Connect these six leads to make a three-phase delta connection.

2. If the rated line voltage of an alternator is 120 volts, how can the alternator be connected for a rated voltage of 208 volts? _____

_____

3. A wye-connected alternator is rated at 20-amperes line current. The internal connections are changed from wye to delta. What is the new line current rating?

_____

4. Connect the three-phase windings in delta in the following diagram.

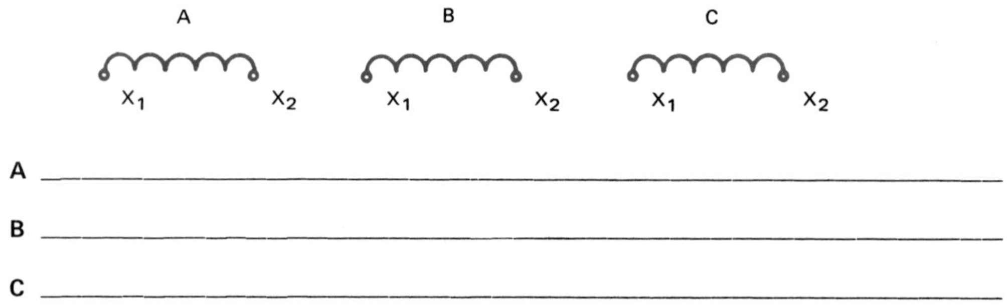

A _____

B _____

C _____

5. What precautions must be taken before closing a delta connection?

_____

_____

_____

_____

_____

_____

_____

_____

Questions 6–10 are based on the diagram below.

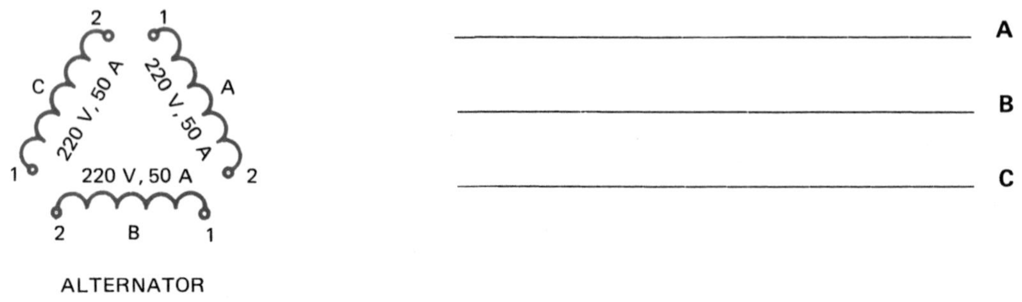

_____ A

_____ B

_____ C

ALTERNATOR

6. Complete the connections for a three-wire delta system._____

_____

_____

_____

_____

_____

7. Determine the rated line current of this three-phase line. _____

_____

_____

8. What is the line voltage?   _____

9. Why is this connection called a delta connection? _____

_____

_____

10. Why is the delta connection limited to three-wire, three-phase transmission circuits?

_____

_____

_____

UNIT 20

SUMMARY REVIEW OF UNITS 17-19

OBJECTIVE

- To give the student an opportunity to evaluate the knowledge and understanding acquired in the study of the previous three units.

Select the correct answers for each of the following statements and place the corresponding letter in the space provided.

1. A polyphase system is a _____
   a. three-phase system.
   b. two-phase system.
   c. six-phase system.
   d. two or more single-phase systems.

2. Three-phase alternator windings are displaced _____
   a. 90 degrees apart.   c. 180 degrees apart.
   b. 120 degrees apart.   d. 360 degrees apart.

3. The two-wattmeter power measurement method _____
   a. cannot be used in an unbalanced three-phase, four-wire system.
   b. is used on single phase only.
   c. can be used in an unbalanced three-phase, four-wire system.
   d. is used on single- and three-phase systems.

4. In a three-phase wye connection, _____
   a. line current = 1.73 × phase current.
   b. line current = phase current.
   c. line voltage = phase voltage.
   d. line voltage × 1.73 = phase voltage.

5. In a three-phase delta connection, _____
   a. line voltage = 1.73 × phase voltage.
   b. line current = phase current.
   c. line voltage × 1.73 = phase voltage.
   d. line voltage = phase voltage.

6. The wye connection is usually wired to a _____
   a. five-wire, three-phase line.
   b. four-wire, three-phase line.
   c. six-wire line.
   d. three-wire, single-phase line.

7. The wye connection is used in
   a. three-phase systems.
   b. two-phase systems.
   c. single-phase systems.
   d. special motor connections.

8. A three-phase, wye-connected generator winding has a phase current rating of 10 amperes. The line current rating is
   a. 10 A.
   b. 30 A.
   c. 17.3 A.
   d. 15 A.

9. The voltage rating of an ac, wye-connected, three-phase generator is 208 volts. The voltage rating of each winding is
   a. 69.3 V.
   b. 208 V.
   c. 120 V.
   d. 110 V.

10. Six leads are brought out of a three-phase transformer. The leads are labeled A_1, A_2, B_1, B_2, C_1, and C_2. The wye connection can be made by connecting leads
    a. A_2, B_2, and C_2.
    b. A_2 to B_1, B_2 to C_1, and C_2 to A_1.
    c. A_1 to B_1 to C_1 and A_2 to B_2 to C_2.
    d. A_2 to B_1 and B_2 to C_1.

11. When the line voltage of a three-phase, four-wire system is 220 volts, the line to ground voltage will be
    a. 110 V.
    b. 220 V.
    c. $\dfrac{220}{1.73}$ V.
    d. $\dfrac{220}{3}$ V.

12. Six leads marked A_1, A_2, B_1, B_2, C_1, and C_2 are brought out of a three-phase transformer bank. If A_2 is connected to B_1, B_2 to C_1, and C_2 to A_1, the transformer windings are connected in
    a. wye.
    b. series.
    c. delta.
    d. parallel.

13. If the phase voltage of a delta-connected generator is 220 volts, the line rated voltage is
    a. 660 V.
    b. 330 V.
    c. 220 V.
    d. 127 V.

14. The delta connection is connected to a
    a. four-wire, three-phase line.
    b. five-wire, three-phase line.
    c. three-wire, three-phase line.
    d. six-wire, three-phase line.

15. An ac generator has delta connections. If each winding is rated at 20 amperes, the line rating is _____
    a. 60 A.   c. 34.6 A.
    b. 20 A.   d. 30 A.

16. A wye-connected ac generator is rated at 208 volts and 25 amperes. The phase winding rating is _____
    a. 208 V, 25 A.   c. 120 V, 14.4 A.
    b. 120 V, 25 A.   d. 208 V, 14.4 A.

17. A delta-connected ac generator is rated at 220 volts and 17.3 amperes. The phase winding rating is _____
    a. 220 V.   c. 381 V.
    b. 127 V.   d. 660 V.

18. A 220-V, 17.3-A delta-connected ac generator is reconnected in wye. The new line voltage rating is _____
    a. 220 V.   c. 381 V.
    b. 127 V.   d. 660 V.

19. When three-phase windings are connected in delta, the coils are connected in _____
    a. an open series circuit.   c. parallel.
    b. a closed series circuit.   d. series parallel.

20. In electrical terminology, the word delta means _____
    a. a deposit at the mouth of a river.
    b. the Greek letter D which is represented by a triangle.
    c. coils joined together at the ends.
    d. coils connected in an open series circuit.

UNIT
21
BASIC PRINCIPLES OF TRANSFORMERS

OBJECTIVES

After studying this unit, the student will be able to

- explain how and why transformers are used for the transmission and distribution of electrical energy.

- describe the basic construction of a transformer.

- distinguish between the primary and secondary windings of a transformer.

- list, in order of sequence, the various steps in the operation of a step-up transformer.

- make use of appropriate information to calculate the voltage ratio, voltages, currents, and efficiency for step-up and step-down transformers.

- explain how the primary loads with the secondary.

It is neither efficient nor economically feasible to generate or transmit large quantities of *direct-current* electrical energy. The invention of the transformer was a milestone in the progress of the electrical industry. The transformer increases or decreases the voltage of large quantities of *alternating-current* energy efficiently, safely, and conveniently. A large power distribution station is shown in figure 21-1.

Fig. 21-1 Substation showing three-phase, oil-filled transformer and three-phase, oil-filled circuit breakers on the right-hand side. Transformer loads are first disconnected by these circuit breakers. *(Courtesy of McGraw-Edison Company, Power Systems Division)*

Large amounts of alternating-current energy may be generated at a convenient voltage, using steam, nuclear, or water power. Transformers are used first to increase this energy to a high voltage for transmission over many miles of transmission wires, and then to decrease this voltage to values which are convenient and safe for use by the consumer.

ELEMENTS OF TRANSFORMERS

A transformer consists of two or more conductor windings placed on the same iron core magnetic path, as shown in figure 21-2.

Laminated Core

The iron core of a transformer is made up of sheets of rolled iron. This iron is treated so that it has a high magnetic conducting quality (high permeability) throughout the length of the core. *Permeability* is the term used to express the ease with which a material will conduct magnetic lines of force. The iron also has a high ohmic resistance across the plates (through the thickness of the core). It is necessary to laminate the iron sheets (figure 21-3) to reduce hysteresis and eddy currents which cause heating of the core.

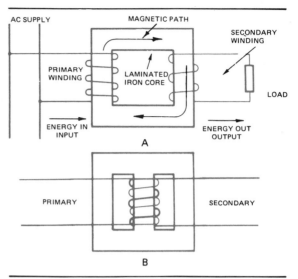

Fig. 21-2 Parts of a transformer

Fig. 21-3 Stacking the iron core of a large three-phase, oil-filled power transformer *(Courtesy of McGraw-Edison Company, Power Systems Division)*

Windings

A transformer has two windings: the primary winding and the secondary winding. The *primary winding* is the coil which receives the energy. It is formed, wound and fitted over the iron core. The *secondary winding* is the coil which discharges the energy at a transformed or changed voltage — increased or decreased.

With reference to the primary winding, when the secondary voltage is lower, the transformer is called a *step-down* transformer. When the secondary voltage is higher, the transformer is called a *step-up* transformer.

The secondary voltage is dependent upon

- the voltage of the primary,
- the number of turns on the primary winding, and
- the number of turns on the secondary winding.

Certain types of core-type transformers have the primary and secondary wire coils wound on separate legs of the core, (see figure 21-2A). The primary and secondary wire coils can also be wound on top of one another, as shown in figure 21-2B. Winding in this manner improves transformer efficiency and conserves energy. When stating the transformer ratio, the primary is the first factor of the ratio. This tells which winding, high or low, is connected to the power source.

CONSTRUCTION OF TRANSFORMERS

Three major types of construction for transformer cores are: core type, shell type, and cross or H type (figure 21-4).

Core Type

In a core-type transformer, the primary winding is on one leg of the transformer and the secondary winding is on the other leg. The most efficient type of core construction is the shell type in which the core is surrounded by a shell of iron.

Shell Type

The shell-type or double window-type core transformer (figure 21-2B), is probably used most frequently in electrical work. In terms of energy conservation, this transformer design operates at 98 percent or higher efficiency.

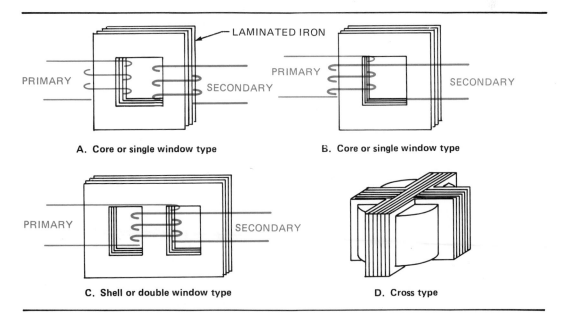

Fig. 21-4 Major construction types for transformer cores

Cross or H Type

The cross or H type of core is also called the modified shell type. The coils are surrounded by four core legs. The cross-type is really a combination of two shell cores set at right angles to each other. The windings are located over the center core which is four times the area of each of the outside legs. This type of core is very compact and can be cooled easily. It is used for large power transformers where voltage drop and cost must be kept to a minimum. These units are usually immersed in oil for high insulation properties and effective cooling. Another method of cooling the transformers is by forced air. Transformers should never be immersed in water for cooling. Accidental flooding, such as in underground transformer vaults, should be pumped.

ELEMENTARY PRINCIPLES OF TRANSFORMER OPERATION

According to Lenz's Law, a voltage is induced in a coil whenever the coil circuit is opened or closed. This induced voltage is always in such a direction as to oppose the force producing it. Called *induction,* this action is illustrated by arranging two loops of wire, as shown in figure 21-5.

Note in figure 21-5 the progressive enlargement of the magnetic field about one side of each loop as the current builds up. The strength of the magnetic field increases as the electrical current through the conductor increases from the power source. Figure 21-5 also shows the field pattern during the period that the current decreases when the circuit is opened.

As the current builds up to its maximum value, the circular magnetic lines around the wire move outward from the wire. This outward movement of magnetic lines of

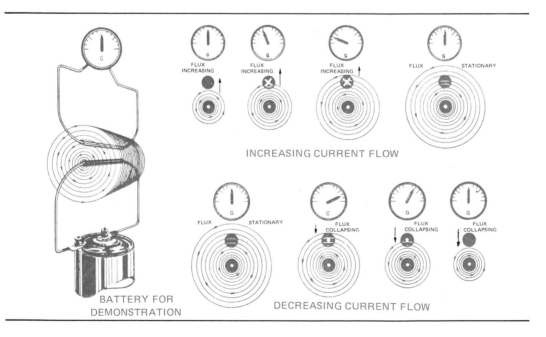

Fig. 21-5 Magnetic induction

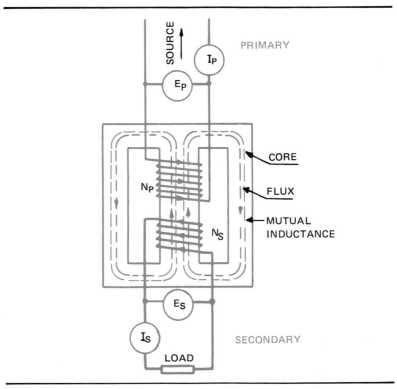

Fig. 21-6 Single-phase transformer showing mutual inductance of two coils

force cuts across the conductor of the second loop. As a result, an emf is induced and current circulates in the loop, as indicated on the galvanometer located above the conductor.

When the current reaches its steady state in the first circuit, the flux is stationary and no voltage is induced in the circuit. The galvanometer indicates zero current.

When the battery circuit is opened, current falls to zero and the flux collapses. The collapsing flux cuts through the second circuit and again induces an emf. The second induced current has a direction opposite to that of the first induced current, as indicated by the galvanometer needle. The final stage shows a steady state with no field and no induced current. This action is automatic with ac applied.

The loops of wire may be replaced by two concentric coils (loops with many turns) to form a transformer. Figure 21-6 shows a transformer which has a primary winding, an iron core and secondary winding. When a changing or alternating current is impressed on the primary winding, the changing primary current produces a changing magnetic field in the iron core. This changing field cuts through the secondary coil and thus induces a voltage, depending on the number of conductors in the secondary coil cut by the magnetic lines. This is called mutual inductance. Commercial transformers generally have fixed cores which provide complete magnetic circuits for efficient operation.

VOLTAGE RATIO

According to Lenz's law, one volt is induced when 100,000,000 magnetic lines of force are cut in one second. The primary winding of a transformer supplies the magnetic field for the core. The secondary winding, when placed directly over the same core, supplies the load with an induced voltage which is proportional to the number of con-ductors cut by the flux of the core.

The shell-type transformer shown in figure 21-6 is designed to reduce the voltage of the power supply.

In figure 21-6,

$$N_p \ = \ \text{number of turns in the primary winding}$$

$$N_s \ = \ \text{number of turns in the secondary winding}$$

$$I_p \ = \ \text{current in the primary winding}$$

$$I_s \ = \ \text{current in the secondary winding}$$

$$\text{Assume that } N_p \ = \ 100 \text{ turns}$$

$$N_s \ = \ \ 50 \text{ turns}$$

$$E_{supply} \ = \ 100 \text{ volts, 60 hertz}$$

The alternating supply voltage produces a current in the primary which magnetizes the core with an alternating flux. According to Lenz's Law, a counter emf is induced in the primary winding. This counter emf is called self-inductance and opposes the im-pressed voltage. Since the secondary winding is on the same core as the primary wind-ing, only 50 volts is induced in the secondary because only half as many conductors are cut by the magnetic field.

At no-load conditions, the following ratio is true:

$$\frac{N_p(100)}{N_s\ (50)} \ = \ \frac{E_p(100)}{E_s\ (50)} \ ; \ \frac{2}{1} \ = \ \frac{2}{1}$$

Therefore, the ratio of 2 to 1 indicates that the transformer is a step-down transformer which will reduce the voltage of the power supply. Transformers either step up or step down the supply voltage.

Refer to figure 21-7 for the following example. The primary winding of a trans-former has 100 turns, and the secondary has 400 turns. An emf of 110 volts is applied to the primary. What is the voltage at the secondary and what is the ratio of the transformer?

$$\frac{E_p}{E_s} \ = \ \frac{N_p}{N_s}$$

$$\frac{110}{E_s} \ = \ \frac{100}{400}$$

$$100 \ E_s \ = \ 44,000$$

$$E_s \ = \ \frac{44,000}{100} \ = \ 440 \text{ volts}$$

This transformer has a $\dfrac{440}{110} \ = \ \dfrac{4}{1}$ step-up ratio.

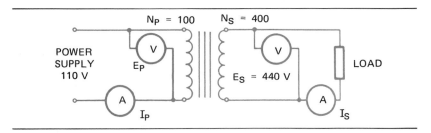

Fig. 21-7 Elementary diagram of a transformer

CURRENT RATIO

If the load current of the transformer shown in figure 21-7 is 12 amperes, the primary current must be such that the product of the number of turns and the value of the current (ampere-turns primary) equal the value of the ampere-turns secondary.

$$N_p I_p = N_s I_s \text{ or } \frac{N_p}{N_s} = \frac{I_s}{I_p}$$

$$\frac{100}{400} = \frac{12}{I_p}$$

$$100\, I_p = 4{,}800$$

$$I_p = 48 \text{ amperes}$$

Check of Solution for Current

$$N_p I_p = N_s I_s; 100 \times 48 = 400 \times 12; \; 4{,}800 = 4{,}800$$

The current ratio is an inverse ratio; that is, the greater the number of turns, the less the current for a given load. Practical estimates of primary or secondary currents are made by assuming that transformers are 100 percent efficient.

For example, assume that

$$\text{Watts input} = \text{Watts output}$$

or

$$\text{Primary watts} = \text{Secondary watts}$$

Therefore, for a 1,000-watt, 100/200-volt step-up transformer:

$$I_s = \frac{1{,}000 \text{ W}}{200 \text{ V}} = 5 \text{ amperes}$$

$$I_p = \frac{1{,}000 \text{ W}}{100 \text{ V}} = 10 \text{ amperes}$$

Obviously, the greater the current the larger size the wire leads are on the transformer. From this information we can determine the high and low voltage sides.

Higher voltage = lower current (p = ie) smaller wire size

Lower voltage = higher current (P = IE) larger size wire

Example: A machine tool being relocated has a control transformer disconnected. The nameplate is illegible due to corrosion. The motor power circuit is 480 volts. The controller is a separate 120 volts. Which is the

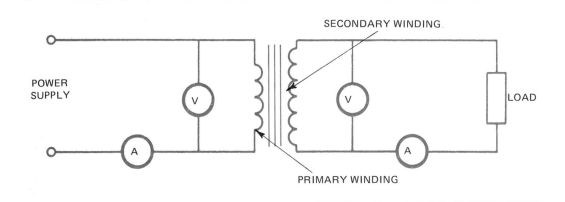

Fig. 21-8 Schematic diagram of a step-up transformer

primary and secondary of the control transformer? The higher voltage has the smaller wire size. Therefore, this is to be connected to the 480 volts.

The use of an ohmmeter can also tell us which winding has the greater resistance. By measuring each winding, we find that the greater the resistance, the greater is the voltage connection. Remember, the term "primary" refers to the supply side of the transformer. The term "secondary" refers to the load side (figure 21-8).

SCHEMATIC DIAGRAM OR SYMBOL

A step-up transformer is usually shown in schematic form, as illustrated in figure 21-8. The ratio of turns, primary to secondary, is not directly shown. This is usually shown as a step-up or step-down symbol representation.

PRIMARY LOADING WITH SECONDARY LOADING

The current in the secondary controls the current in the primary. When the secondary circuit is closed by placing a load across it, the secondary emf causes a current to flow. This builds up a magnetic field in opposition to the primary field. This opposing, or demagnetizing, action reduces the effective field of the primary cemf, which in turn reduces the primary cemf, thereby permitting more current to flow in the primary. The greater the current flow in the secondary, the greater is the field produced by the secondary. This results in a reduced primary field; hence, a reduced primary cemf is produced. This condition permits greater current flow in the primary. This entire process, which is instantaneous, will repeat itself whenever there is any change in the value of the current in either the primary or the secondary. A transformer adjusts itself readily to any normal change in secondary impedance. However, if a direct short is placed across the secondary, the abnormally great amount of current flowing causes the primary current to rise in a like manner, resulting in damage to, or complete burn-out of, the transformer, if it is not protected properly.

EFFICIENCY

The efficiency of all machinery is the ratio of the output to the input.

$$\text{Efficiency} = \frac{\text{output}}{\text{input}}$$

In general, transformer efficiency is about 97 percent. Only three percent of the total voltage at the secondary winding is lost through the transformation. The loss in voltage is due to *core losses* and *copper losses*.

The core loss is the result of hysteresis (magnetic friction) and eddy currents (induced currents) in the iron core.

The copper loss is power lost in the copper wire of the windings ($I^2 R$). Therefore, taking these losses into consideration,

$$\text{Efficiency} = \frac{\text{Watts output (secondary)}}{\text{Watts input (primary)}}$$

where Watts input = Watts output + losses

ACHIEVEMENT REVIEW

A.   Select the correct answer for each of the following statements and place the corresponding letter in the space provided.

1. When the primary winding has more turns than the secondary, the voltage in the secondary winding is   _____
    a. increased.   c. decreased.
    b. doubled.   d. halved.

2. In the coils of a transformer, the motion of the flux is caused by the   _____
    a. direct current.   c. moving secondary.
    b. rotating primary.   d. alternating current.

3. Energy is transferred from the primary to the secondary coils without a change in   _____
    a. frequency.   c. current.
    b. voltage.   d. ampere-turns.

4. Transformer efficiency averages   _____
    a. 79 percent.   c. 50 percent.
    b. 97 percent.   d. 100 percent.

5. A transformer has a primary coil rated at 150 volts and a secondary winding rated at 300 volts. The primary winding has 500 turns. How many turns does the secondary winding have?   _____
    a. 250   c. 1,000
    b. 2,500   d. 10,000

6. A control transformer is a step-down-type transformer. Com-
    pared to the secondary winding, the primary winding is   _____
    a. larger in wire size.
    b. smaller in wire size.
    c. the same size as the secondary.
    d. connected to the load.

7. The current in the secondary winding   _____
    a. is higher than the current in the primary.
    b. is lower than the current in the primary.
    c. controls the current in the secondary.
    d. controls the current in the primary.

B.   Solve the following problems.

8. A 110/220-volt step-up transformer has 100 primary turns.
    How many turns does the secondary winding have?   _____

9. A transformer has 100 primary turns and 50 secondary turns.
    The current in the secondary winding is 20 amperes. What is
    the current in the primary winding?   _____

10. What is the ratio of a transformer that has a secondary volt-
    age of 120 volts when connected to a 2,400-volt supply?   _____

11. A 7,200/240-volt step-down transformer has 1,950 primary
    turns.   Determine the number of turns in the secondary
    winding.   _____

12. A 2,400/240-volt step-down transformer has a current of 9
    amperes in its primary and 85 amperes in its secondary. Deter-
    mine the efficiency of the transformer.   _____

SINGLE-PHASE
TRANSFORMERS

OBJECTIVES

After studying this unit, the student will be able to

- describe a single-phase, double-wound transformer, including its primary applications.
- diagram the series and parallel methods of coil connection for a double-wound transformer and for dual-voltage connections, primary and secondary.
- define what is meant by subtractive polarity and diagram the connections and markings for this polarity.
- define what is meant by additive polarity and diagram the connections and markings for this polarity.
- list the steps in the ac polarity test for a single-phase transformer.
- demonstrate good electrical safety practices.
- describe an autotransformer, including its primary applications.
- identify primary taps.

A single-phase transformer usually has a core and at least two coils. The single-phase autotransformer has only one coil. The specifications for single-phase transformers vary greatly and the applications of these transformers are unlimited.

THE DOUBLE-WOUND TRANSFORMER (ISOLATING AND INSULATING)

The double-wound transformer has a primary winding and a secondary winding. These windings are independently isolated and insulated from each other. A *shielded winding* transformer, on the other hand, is designed with a metallic shield between the primary and secondary windings, providing a safety factor by grounding. This prevents accidental contact between the windings under faulty conditions. The illustrations in unit 21 show a double-wound transformer. The coils of double-wound transformers may be connected in several different arrangements.

Figure 22-1 shows two popular arrangements of single-phase transformer windings. Two single coils (figure 22-1A), are used for specific step-down or step-up applications, including bell ringing transformers, neon transformers, and component transformers for commercial equipment, such as automatic machines, switchgear, and other devices. Multiple coil primary and secondary windings (figure 22-1B), are used in distribution transformers where dual voltage ratings are desired. Arrangements for voltage ratings of 2,400/120/240 or 220/440/110/220 are common.

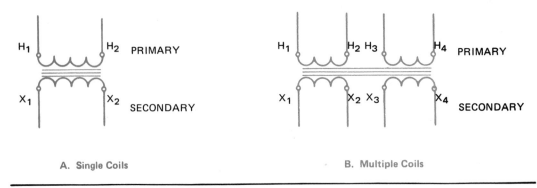

A. Single Coils   B. Multiple Coils

Fig. 22-1 Coil arrangements for single-phase transformers

Double-wound transformers separate or insulate the high transmission voltages from the typical consumer voltages of 110/220/440. The National Electrical Code requires the use of this type of transformer in all distribution circuits with the exception of those circuits assigned to autotransformers. Here, as in the Code, the voltage considered shall be that at which the circuit operates, except for the examples given.

Polarity

A 440/110/220 transformer may be connected for two ratios:

440/110 or 440/220

To obtain the 440/110 ratio, the secondary coils are connected in parallel; the 440/220 ratio is achieved by connecting the secondary coils in series. To complete these connections, the polarity of the leads must be determined.

Figure 22-2 shows how the transformer series and parallel coil connections are made. Note that instead of polarity indications such as (+, −) the coil leads are identified here by S (start) and F (finish), or by 1 (start) and 2 (finish) as in H_1, H_2 and X_1, X_2.

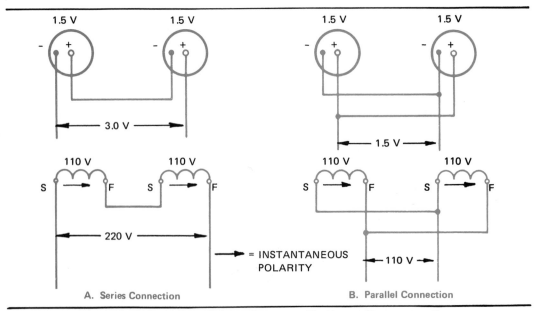

A. Series Connection   B. Parallel Connection

Fig. 22-2 Series and parallel transformer and battery voltage connections

The beginning or ending of a transformer coil is usually indicated by a tab placed on the lead by the technician in charge of the winding process. When the transformer is assembled, other markings often replace the original ones. Before final inspection, a polarity test must be made to be certain that the leads are marked correctly.

IDENTIFYING AN UNMARKED TRANSFORMER

Installed transformers often have missing or disfigured tabs. Every time a transformer is to be reconnected following repairs, or must be reconnected for other reasons, the polarity of the leads must be checked.

Figures 22-3 and 22-4 illustrate two systems of marking polarity. In conventional usage, polarity refers to the induced voltage vector relationships of the transformer leads as they are brought outside of the tank. The American National Standards Institute has standardized the location of these leads to obtain additive and subtractive polarity conditions. All high-voltage leads brought outside the case are marked H_1, H_2, and so forth, while the low-voltage leads are marked X_1, X_2. The H_1 lead is located on the left side when facing the low-voltage leads. H_1 and X_1 are both positive at the same instant of time.

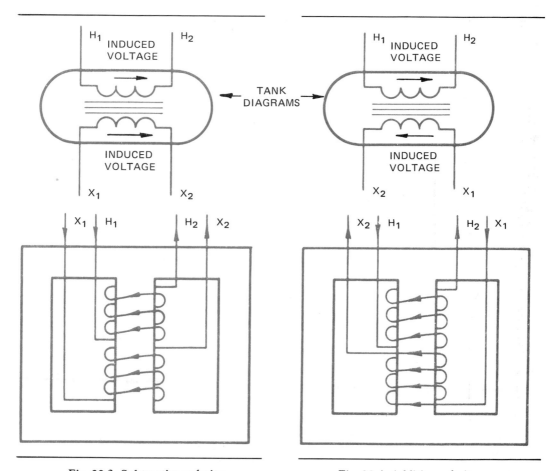

Fig. 22-3 Subtractive polarity   Fig. 22-4 Additive polarity

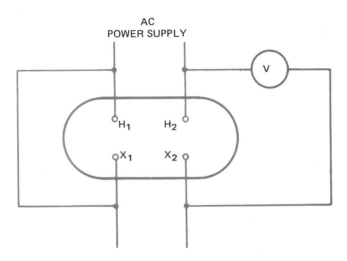

Fig. 22-5 Ac polarity test of a single-phase transformer

Subtractive Polarity. The tank diagram in figure 22-3 shows the relationship of the induced voltages in the primary and secondary windings for the subtractive polarity condition. Transformers connected in this manner have the H_1 and X_1 leads located directly opposite each other. If H_1 and X_1 are connected together (as shown in figure 22-5), the voltage measured between H_2 and X_2 is less than the primary voltage. The induced voltages oppose each other and thus cause the secondary induced voltage to be subtracted from the primary voltage.

Additive Polarity. The tank diagram in figure 22-4 shows the voltage relationship of the induced voltages for the additive polarity connection. When H_1 and X_2 are connected, the voltage across H_2 and X_1 is greater than the primary voltage. The induced voltages add up to the sum of the primary and secondary voltages.

Transformers which are rated up to 200 kVA and have the value of the high-voltage winding equal to 8,660 volts or less will be additive. All other transformers will be subtractive.

Test for Polarity. Transformer coils often must be connected in series or parallel before the lead markings are definitely determined. For these situations, the polarity of a transformer or any secondary coil can be found by making the connections shown in figure 22-5. Connect the adjacent left-hand, high-voltage and low-voltage outlet leads facing the low-voltage side of the transformer. Apply a low-voltage supply to the primary and note the voltage between the adjacent right-hand, high- and low-voltage terminals.

- For *subtractive* polarity, the voltmeter reading (V) is less than the applied voltage. The voltage is the difference between the primary and secondary voltages, $E_p - E_s$.
- For *additive* polarity, the voltmeter reading (V) is greater than the applied voltage. The voltage is the sum of the primary and secondary voltages, $E_p + E_s$.

If the test shown in figure 22-5 indicates additive polarity, the secondary leads inside the tank must be reversed at the bottom of the bushings to obtain a true subtractive polarity. If the transformer requires an additive polarity and the test indicates the same, reverse the secondary lead markers so that X_2 is located opposite H_1. Most transformers are connected in additive polarity.

In all transformers, the H terminals are always the high voltage terminals. The X terminals are always the low voltage terminals. Transformers permitting, either the H or X terminals can be designated as the primary or the secondary, depending upon which is the source and which is the load.

SINGLE-PHASE TRANSFORMER CONNECTIONS

Series Connection

If a 460/115/230-volt single-phase transformer is to be connected to obtain 440/220 volts, the two secondary coils must be connected in series. The beginning and ending of each coil must be joined, as shown in figure 22-6A. The "start" of each coil is identified by an odd-numbered subscript.

Note: If the voltage is zero across $X_1 - X_4$ after the series connections are complete, the coils are opposing each other (the polarity of one coil is reversed). To correct this situation, reverse one coil, then reconnect and recheck the polarity.

Parallel Connection

To obtain 460/115 volts, the two secondary coils must be connected in parallel as shown in figure 22-6B. The polarity of each coil must be correct before making this connection. The parallel connection of two coils of opposite polarity will result in a short circuit and internal damage to the transformer.

Note: An indirect polarity check can be made by completing the series connection and noting the total voltage. As noted above, zero voltage indicates opposite polarities.

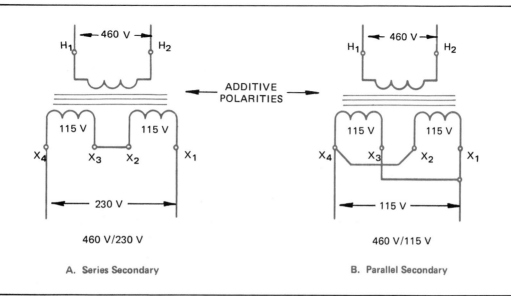

Fig. 22-6 Single-phase transformer connections

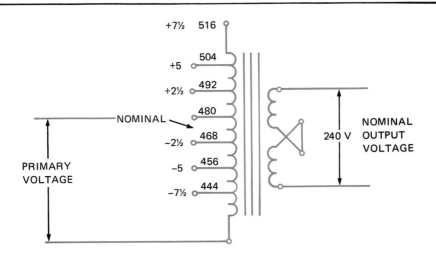

Fig. 22-7 Primary terminal taps

Reverse one coil to remedy the condition and then recheck overall polarity. **Retag** leads if necessary.

PARALLEL OPERATION OF SINGLE-PHASE TRANSFORMERS

Single-phase distribution transformers can be connected in parallel only if the voltage and percent impedance ratings of the transformers are identical. This information is found on the nameplates of large size transformers. It is recommended that this rule be followed when making permanent parallel connections of all transformers.

TRANSFORMER PRIMARY TAPS

Taps are nothing more than alternative terminals which can be connected to more closely match the supply, primary voltage. These taps are arranged in increments of 2½ percent or 5 percent of the primary nominal voltage rating of the transformer (figure 22-7). This provides a job site adjustment to ensure that the primary of the transformer matches the supply voltage. The secondary will then produce the desired secondary voltage.

The voltage received from the power utility may be low or high. Since the transformer is a fixed voltage device, the output voltage is always in direct proportion to the input voltage. If the ratio is 2:1 and the supply voltage is 480 volts, the output will be 240 volts. If the primary voltage is 438 volts, the secondary will be only 219 volts.

High and low voltages can have serious effects on different connected loads. Therefore, care must be taken to deliver a voltage as close as possible to the desired primary — so that the secondary voltage will match the equipment nameplate voltages. Consistently high and low voltage problems can be solved by connecting the proper primary taps (figure 22-8). If the voltage fluctuates consistently, this can be neglected.

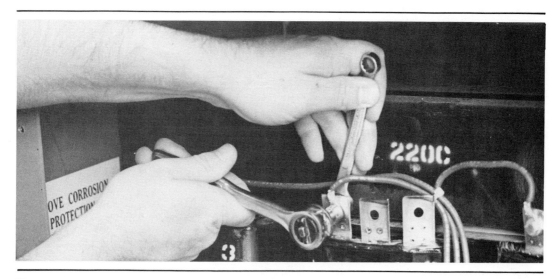

Fig. 22-8   Taps are easily changed.   The diagrammatic nameplate shows exact tap connections for high and low voltage conditions. *(Courtesy of Hevi-Duty Electric, A Unit of General Signal)*

REGULATION

A slight *voltage drop* at the secondary terminals from *no load* to *full load* is called regulation; these are resistance and reactance drops in the windings as losses.   Regulation is expressed as a percentage.   Regulation of constant potential transformers is about 1 percent to 5 percent.   Secondary terminals:

$$\% \text{ E regulation} = \frac{\text{no load E} - \text{full load E}}{\text{full load E}} \times 100$$

Example: The secondary voltage of a transformer rises from 220 to 228 volts when the rated load is removed.   What is the regulation of the transformer?

$$\% \text{ regulation} = \frac{228 - 220}{220} = .036 \times 100 = 3.6\%$$

AUTOTRANSFORMER

Transformers having only one winding are called autotransformers.   An autotransformer is a transformer in which a part of the winding is common to both primary and secondary circuits.   This is the most efficient type of transformer since a portion of the one winding carries the difference between the primary and secondary currents. Figure 22-9 shows the current distribution in an autotransformer used in a typical lighting application.   The disadvantage of an autotransformer is the fact that the use of only one winding makes it impossible to insulate the low-voltage section from the high-voltage distribution line.   If the low-voltage winding opens when stepping down the voltage, the full line voltage appears across the load.   According to the National Electrical Code, the use of autotransformers is limited to certain situations, such as where

   a. the system supplied contains an identified grounded conductor which is solidly connected to a similar identified grounded conductor of the system supplying the autotransformer;

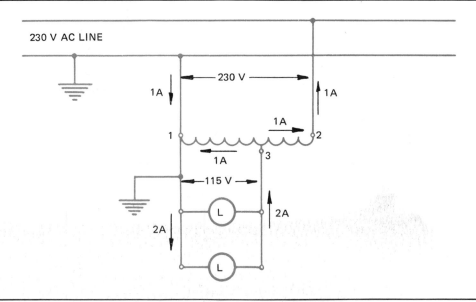

Fig. 22-9 Autotransformer used for lighting

b. an induction motor is to be started or controlled (figure 22-10);

c. a dimming action is required, as in theater lighting;

d. the autotransformer is to be a part of a ballast for supplying lighting units;

e. to boost or to buck a voltage under certain conditions.

Fig. 22-10 An autotransformer assembly used for starting induction motors at reduced voltages *(Courtesy of Hevi-Duty Electric, A Unit of General Signal)*

DRY AND LIQUID-FILLED TRANSFORMERS

Dry transformers are used extensively for indoor installations. These transformers are cooled and insulated by air and are not encased in heavy tanks, such as those required for liquid-filled transformers. Dry transformers are used for bell ringing circuits, current and potential transformers, welding transformers, and almost all transformers used on portable or small industrial equipment.

Liquid-filled transformers consist of the core and coils immersed in a tank of oil or other insulating liquid. Oil insulation is approximately fifteen times more effective than air insulation. Most distribution transformers designed for outdoor installation are liquid filled.

METHODS OF COOLING

The method selected to cool a transformer must not only maintain a sufficiently low average temperature, but must also prevent an excessive temperature rise in any portion of the transformer. In other words, the cooling medium must prevent the formation of "hot spots." For this reason, the working parts of the transformer are usually immersed in a high grade of insulating oil. The oil must be free of any moisture so, if necessary, the oil must be filtered to remove moisture. The insulating value of the oil is checked periodically.

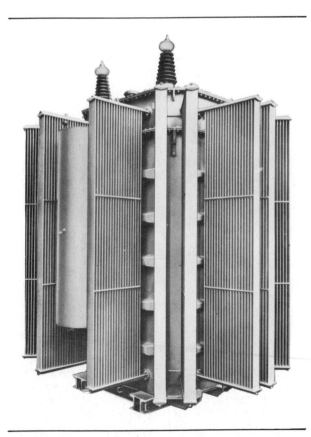

Fig. 22-11 An oil-filled power transformer with radiators

Fig. 22-12 Assembly of a large, three-phase, oil-filled station-class power transformer, such as the one shown in figure 22-11   *(Courtesy of McGraw-Edison Company, Power Systems Division)*

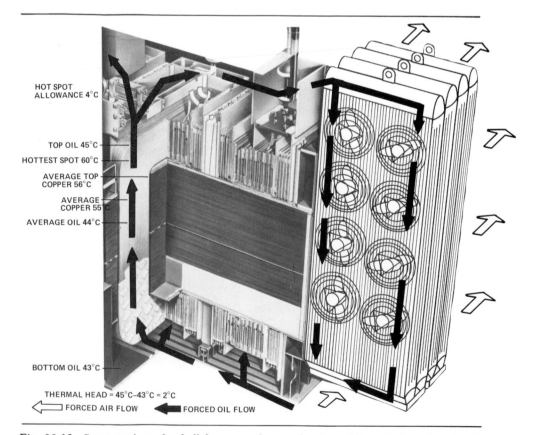

Fig. 22-13 Cross section of a shell form transformer showing oil-forced air cooling (FOA or FOA/FOA with typical temperature rises *(Courtesy of Westinghouse Electric Corporation, Power Transformer Division)*

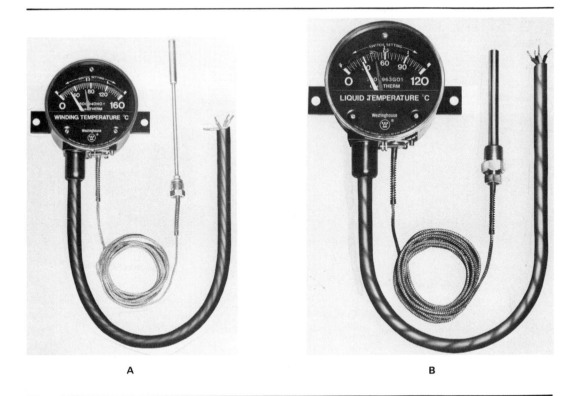

Fig. 22-14   Transformer temperature indicators   (A) Winding temperature indicator   (B) Liquid temperature indicator   *(Courtesy of Westinghouse Electric Corporation, Power Transformer Division)*

Duct lines are arranged within the transformer to provide for the free circulation of oil through the core and coils.   The warmer and thus lighter oil rises to the top of the steel tank in which the transformer core and windings are placed.   The cooler and heavier oil settles to the bottom of the tank.   This natural circulation provides for better cooling.

Forced Cooling

Several methods of removing heat from a transformer involve forced cooling.   Cooling is achieved by using pumps to force the circulation of the oil or liquid, by forcing the circulation of air through the radiators (figure 22-11) or by immersing water-containing coils in the oil.   Cold water circulating in the coils removes the heat stored in the oil.   Forced air movement by the use of fans is a common practice (figure 22-13).   These are generally controlled by thermostats (figure 22-14).

APPLICATION

Single-phase transformers are suitable for use in a wide variety of applications as shown by the examples illustrated by figures 22-15, 22-16, 22-17, and 22-18.

Distribution transformers are usually oil filled and mounted on poles, in vaults, or in manholes.

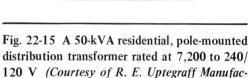

Fig. 22-15 A 50-kVA residential, pole-mounted distribution transformer rated at 7,200 to 240/ 120 V *(Courtesy of R. E. Uptegraff Manufacturing Company)*

Fig. 22-16   Cutaway view of a distribution transformer   *(Courtesy of General Electric Company)*

Compensator starters are tapped autotransformers which are used for starting induction motors.

Instrument transformers such as potential and current transformers are made in indoor, outdoor, and portable styles used for metering.

Welding transformers provide a very low voltage to arc welding electrodes. Movable secondaries provide the varying voltage and current characteristics required.

Constant-current transformers are used for series street lighting where the current must be held constant with a varying voltage. The secondary is movable.

Fig. 22-17 Industrial control transformer *(Courtesy of Hevi-Duty Electric, A Unit of General Signal)*

Fig. 22-18   Power transformer used in radio receiving apparatus *(Courtesy of General Electric Company)*

High voltage test sets include a voltage regulator connected to the primary of a high-voltage, step-up transformer. These sets are used for high potential tests of electrical equipment and distribution lines.

SAFETY PRECAUTIONS

Although there are no moving parts in a transformer, there are some maintenance procedures that must be performed. For a general overhaul of an operating transformer or when an internal inspection is to be made, the transformer must be de-energized. Do not assume either that the transformer is disconnected or rely on someone else to disconnect it; always check the transformer yourself. You must be sure that the fuses are pulled open or out and that the switch or circuit breaker is open on both the primary and secondary sides. After the transformer is disconnected, the windings should be grounded to discharge any capacitive energy stored in the equipment. This step protects you while you are at work also. Grounding is accomplished with a device commonly known as a "short and ground." This is a flexible cable with clamps on both ends. The ground end is clamped first, then, using a hook stick, the other end is touched to the conductor. Do this with each leg on the primary and secondary sides. The phases are then shorted together and grounded for your protection.

The tank pressure should be relieved. This may be done by bleeding a valve or plug above the oil level. Any gas in the tank must be replaced with fresh air before a person enters the tank. The absence of oxygen in a tank will cause asphyxiation quickly and without warning. A second person should be on duty outside the transformer as a safety precaution whenever someone must enter the transformer. All tools should have safety cords attached with the other end tightly secured. All pockets in clothing should be emptied. Nothing must be allowed to fall into the tank. Great care must be exercised to prevent contacting or coming close to the electrical conductors and other live parts of the transformer unless it is known that the transformer has been de-energized. The tank and cooling radiators should not be touched until it is determined that they are adequately grounded (for both new and old installations).

ACHIEVEMENT REVIEW

Select the correct answers for each of the following statements and place the corresponding letter in the space provided.

1. Double-wound transformers contain a minimum of   _____
   a. one main winding.
   b. one main winding with two coils.
   c. a primary and a secondary winding.
   d. a primary and a double-wound secondary.

2. A transformer has subtractive polarity when the   _____
   a. two primary coil voltages oppose each other.
   b. two secondary coils have opposite polarities.
   c. X_1 lead is opposite the H_1 lead.
   d. X_2 lead is opposite the H_1 lead.

3. A transformer has additive polarity when the
   a. two primary coils are in series.
   b. two secondary coils have aiding polarities.
   c. X_1 lead is opposite the H_1 lead.
   d. X_2 lead is opposite the H_1 lead.

4. Polarity should be tested before
   a. energizing a transformer.
   b. checking the ratio.
   c. connecting the coils in series or parallel.
   d. connecting the load to the secondary.

5. A 440/110/220-volt step-down transformer is connected for
   440/220 V. Preliminary tests show that each secondary coil
   has 110 volts but the voltage across $X_1 - X_4$ is zero. The
   probable trouble is that
   a. the voltages in the coils are equal and opposing.
   b. their ratings are equal.
   c. the load will divide in proportion to the capacities.
   d. the voltage drops at full load will be proportional to their
      respective loads.

6. The autotransformer may be used as a
   a. power transformer.
   b. potential transformer.
   c. current transformer.
   d. compensator motor starter.

7. Insulation of transformers may be classed in two groups:
   a. double-wound and autotransformers.
   b. dry and oil-filled types.
   c. core and shell types.
   d. core and cross types.

8. Regarding cooling, transformers may be
   a. air- and oil-cooled.
   b. outdoor- and indoor-cooled.
   c. self- and forced-cooled.
   d. dry- and liquid-cooled.

9. Single-phase, double-wound transformers must be used for
   a. distribution and compensator starters.
   b. instrument and welding transformers.
   c. welding and dimming in theater lighting.
   d. constant current and reduced voltage motor starters.

10. For low voltage, the secondary of a single-phase transformer is connected:

    a. X_1 and X_4 to load, X_3 and X_2 together.
    b. X_1 to X_3 to load, X_2 to X_4 to load.
    c. H_1 to H_4 to load, H_3 to H_2.
    d. H_1 to H_2, X_1 to X_2.

11. A transformer in which part of the secondary is part of the primary is

    a. a series and parallel connection.
    b. a double-wound transformer.
    c. an autotransformer.
    d. an isolating transformer.

12. Parallel operation of single-phase transformers can be accomplished when the

    a. voltage and percentage impedance ratings are identical.
    b. voltage and current ratings are equal.
    c. cooling methods are identical.
    d. primary and secondary voltage ratings are equal.

13. Primary taps are designed to

    a. raise the voltage of the secondary.
    b. drain the oil.
    c. lower the voltage of the secondary.
    d. raise or lower the voltage of the secondary.

14. A slight voltage drop at the secondary terminals from no load to full load is called

    a. reactance.   c. percentage.
    b. regulation.   d. taps.

15. When working in a large transformer, the electrician should

    a. ventilate it first.
    b. short and ground all windings.
    c. secure all tools and empty pockets.
    d. all of these.

16. When preparing to work on an oil-filled transformer,

    a. bleed the tank pressure.
    b. disconnect the supply voltage and load.
    c. disconnect all other connections.
    d. check the disconnect switches yourself.

THE SINGLE-PHASE, THREE-WIRE SECONDARY SYSTEM

OBJECTIVES

After studying this unit, the student will be able to

- diagram the connections for a single-phase, three-wire secondary system.
- list the advantages of a three-wire service.
- describe what occurs when the neutral of a three-wire secondary system opens.
- explain why there is less copper loss for a three-wire system.

Most new homes are wired for three-wire service. Since electric ranges and air conditioners are designed for three-wire operation, any home which is to be provided with these appliances must have three-wire service. The three wires terminate in the residence at the load center panel so that most individual circuits carried through the house are at 115 volts, thus eliminating the dangers associated with 230-volt circuits.

The double-wound transformer is used as the source for three-wire secondary distribution. One of the important advantages of a transformer is its ability to provide a three-wire circuit from the low-voltage secondary. A step-down transformer with a 2,300/230/115-volt rating is commonly used in residential installations.

The advantages of the use of three-wire service in general distribution systems include (1) a reduction in the cost of main feeders and subfeeders, (2) the provision of 115-volt service for normal lighting circuits and 230-volt service for power and motor loads, and (3) the conservation of electrical energy by reducing voltage drop (loss).

Figure 23-1 is a schematic of a typical three-wire system. The secondary coils are connected in series and each coil is rated at 115 volts. The junction N between the two secondary coils is usually grounded. This precaution provides some protection to an individual who may come into contact accidentally with a transformer that has faulty insulation. The line wire carried from this junction to the several loads is known as the *neutral* or *identified conductor.* The neutral wire generally carries less current than wires L_1 and L_2, except when the load is on one side only, that is L_1 to N or L_2 to N. The 230-volt motor load does not affect the current flowing in the neutral wire. The neutral is carried through the system as a solid conductor (not fused or switched). If the neutral opens, and the loads in the 115-volt circuits are greatly unbalanced, then these 115-volt circuits will be subjected to approximately 230 volts. The neutral is designed to carry not only the unbalanced current in the two 115-volt circuits, but also the entire load on any one side should all the load on the other side be cut off completely. This latter situation can occur if a fuse or

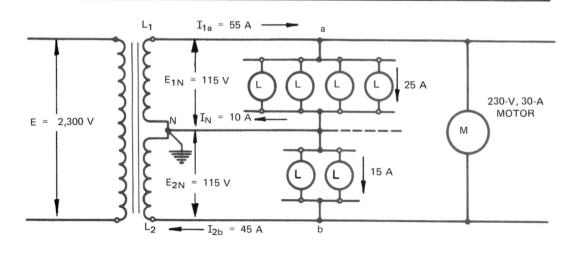

Fig. 23-1 Schematic diagram of a three-wire system supplied from a single-phase transformer

circuit breaker suddenly opens either line. Figure 23-1 shows the current distribution for the loads indicated.

OPEN NEUTRAL

As an example to show what occurs when the neutral of a three-wire system opens, assume that the lighting load in figure 23-1 is a pure resistive load.

Thus, the group of four lamps has a resistance of: $\dfrac{E}{I} = \dfrac{115}{25} = 4.6$ ohms, and the group of two lamps has a resistance of: $\dfrac{E}{I} = \dfrac{115}{15} = 7.666$ ohms.

With the neutral open, these two groups combine as a series circuit with a resistance of 12.266 ohms connected across 230 volts. The current flow through this series circuit is:

$$\frac{E}{R} = \frac{230}{12.266} = 18.75 \text{ amperes}$$

Then, according to the laws for a series circuit, the voltage across the 7.666-ohm group (two lamps) is equal to:

$$I \times R = 18.75 \times 7.666 = 143.74 \text{ volts}$$

and the voltage across the 4.6-ohm group (four lamps) is equal to:

$$I \times R = 18.75 \times 4.6 = 86.25 \text{ volts}$$

(Remember that in a series circuit, the highest voltage appears across the highest value of resistance.) The lamps would probably burn out with this open neutral.

Sample Problem

Referring to figure 23-1, assume that the upper 115-volt load is 25 amperes, the lower load is 15 amperes, and the motor load is 30 amperes. If the power factor in all cases is unity (1), calculate the current

1. in line 1-a.
2. in line 2-b, and
3. in the neutral line N.

In addition, determine the power delivered

4. by transformer coil 1-N,
5. by transformer coil N-2, and
6. by the primary coil.

Finally, calculate the current

7. in the primary coil.

Solution

1. I_{1-a} = 25 + 30 = 55 amperes
2. I_{2-b} = 15 + 30 = 45 amperes
3. I_N = 25 - 15 = 10 amperes
4. P_{1-N} = 55 X 115 = 6,325 watts
5. P_{N-2} = 45 X 115 = 5,175 watts
6. P_{pri} = 6,325 + 5,175 = 11,500 watts
7. I_{pri} = 11,500/2,300 = 5 amperes

The distribution transformers used in industrial plants or network substations for three-wire secondary systems are usually mounted on poles (figure 23-2) or in transformer vaults. This type of transformer is equipped with three low-voltage bushings and the series connection is made inside the tank. The lower lines constitute the secondary three-wire systems.

ECONOMICS OF THE THREE-WIRE SYSTEM
FOR FEEDERS AND BRANCH CIRCUITS

Using the three-wire system of the previous problem as an example, the total load transmitted over the three wires is 11,500 W or 11.5 kW at a power factor of 100 percent. It is assumed that the motor load is provided with power factor correction. If single conductor-type TW wire is used from the transformer to the load, the following sizes are required.

Line 1 (55 amperes): No. 6 TW

Neutral (0.70 X 55 = 38.5): No. 8 TW

Line 2 (55 amperes): No. 6 TW

Although No. 8 TW wire is the actual size permitted for the neutral, a substitution of a No. 6 TW wire can be made so that three No. 6 lines are provided to simplify the installation.

If a two-wire distribution system is used for the same load, the total current is 11,500/115 = 100 amperes. Two No. 1 lines are required. If the transmission distance

Fig. 23-2 Pole-type transformer with a low-voltage, three-wire distribution system at the bottom *(Courtesy of General Electric Company)*

is 100 feet, then a comparison can be made of the weights of copper wire required for the two systems.

Three-Wire System

For a No. 6 TW line, the weight per 100 feet = 11.5 lb. Therefore, for 3 No. 6 TW lines, the total weight = 3 × 11.5 = 34.5 lb.

Two-Wire System

For a No. 1 line, the weight per 100 feet = 33 lb.

For 2 No. 1 lines, the total weight = 2 × 33 = 66 lb.

Therefore, for the same load, the three-wire system uses less copper (66 − 34.5 = 31.5 pounds less) than the two-wire system.

A similar conclusion can be reached by consulting a manufacturer's price list and noting the lower prices for smaller size conductors. The copper losses in the line are also considerably less for a three-wire system for several reasons: the motor power is transmitted at a higher voltage and, therefore, less current is required for a given load; the neutral carries no current when the two lighting circuits are balanced; the copper losses are much less because less wire and current are required. These line losses are of two types: (1) voltage drop (IR), and (2) wattage loss (I^2R).

ACHIEVEMENT REVIEW

1. Cite two reasons why power companies must supply three-wire service to residential occupancies. _____

    _____

    _____

2. How are the two secondary coils of a distribution transformer connected for three-wire service? _____

    _____

3. What are three advantages of a three-wire service as compared to a conventional two-wire service? _____

    _____

    _____

    _____

4. Why must the neutral line be left unfused? _____

    _____

    _____

5. How many circuits are provided in a three-wire secondary system? _____

    _____

6. What is the voltage rating of each circuit in question 5? _____

    _____

    _____

    _____

Questions 7–9 are based on the following problem: a three-wire system has one lighting load of 40 amperes, one lighting load of 20 amperes, and a 230-volt motor load of 30 amperes.

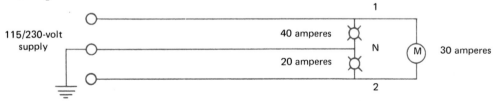

7. What is the current load in lines 1 and 2?

_____   _____

            (1)   (2)

8. What is the current in N?   _____

9. If the neutral is open, indicate the voltages of the lighting circuits. Show the work.

1-N _____   2-N _____

10. A three-wire, 120/240-volt circuit supplies the following:

      One 120-volt, 10-watt lamp to line 1 and neutral, and
      One 120-volt, 120-watt TV set to line 2 and neutral. (See diagram below.)

If the neutral opens while both the lamp and the TV set are operating, what will be the voltage at the lamp and the voltage at the TV set? (Assume power factor of unity.)

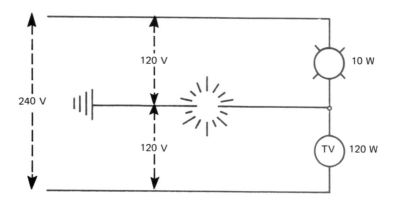

         (lamp voltage)   _____

         (TV voltage)   _____

11. A manufacturer uses motors larger than 1 hp on 120 volts. Why does this require a three-wire secondary system? _____

_____

_____

12. Theoretically, how much horsepower is on the following unbalanced lines?
    L₁ _____   L₂ _____   Neutral _____

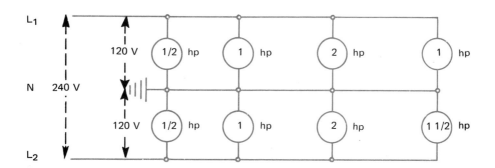

SINGLE-PHASE TRANSFORMERS CONNECTED IN DELTA

UNIT 24

OBJECTIVES

After studying this unit, the student will be able to

- explain, with the aid of diagrams, how single-phase transformers are connected in a three-phase, closed-delta-delta arrangement.

- describe the relationships between the voltages across each coil and across the three-phase lines for both the input (primary) and output (secondary) of a delta-delta transformer bank.

- list the steps in the procedure for checking the proper connection of the secondary coils in the closed-delta arrangement; include typical voltage readings.

- describe how a delta-delta-connected transformer bank can provide both a 240-volt, three-phase load and a 120/240-volt, single-phase, three-wire load.

- describe, using diagrams, the open-delta connection and its use.

- identify primary taps for three-phase connection.

Most electrical energy is generated by three-phase alternating-current generators. Three-phase systems are used for the transmission and distribution of the generated electrical energy. The voltage on three-phase systems often must be transformed, either from a higher value to a lower value, or from a lower value to a higher value.

Voltage transformation on three-phase systems is usually obtained with the use of three single-phase transformers (figure 24-1). These transformers can be connected in several ways to obtain the desired voltage values.

A common connection pattern that the electrician is often required to use for the three single-phase transformers is the closed-delta connection.

Another connection pattern which is commonly used is the open-delta or V connection which requires only two transformers to transform voltage on a three-phase system.

CLOSED-DELTA CONNECTION

When three single-phase coils are connected so that each coil end is connected to the beginning of another coil, a simple closed-delta system is formed (figure 24-2).

When the three coils are marked Coil A, Coil B, and Coil C, the end of each of the three coils is marked with the letter O. The beginnings of the coils are marked A, B, and C. Note that each coil end is connected to another coil beginning. Each of the three junction points ties to a line lead feeding a three-phase system.

Fig. 24-1 Three large single-phase, station-class, oil-filled power transformers
(Courtesy of McGraw-Edison Company, Power Systems Division)

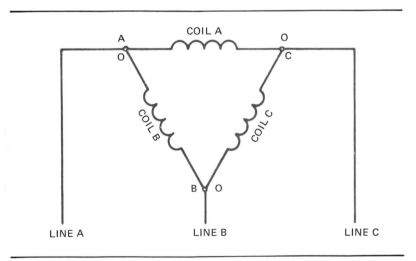

Fig. 24-2 Simple delta connection

If three single-phase transformers are to be used to step down 2,400 volts, three phase, to 240 volts, three phase, a closed-delta connection is used. Each of the three transformers is rated at 2,400 volts on the high-voltage side and 240 volts on the low-voltage side (figure 24-3).

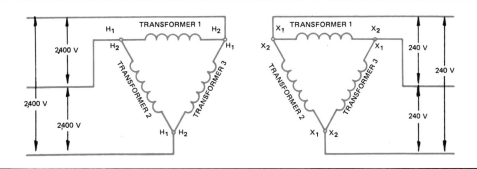

Fig. 24-3 Elementary diagram of delta-delta transformer connections

CONNECTING THE DELTA

The transformer leads on the high-voltage input or primary side of each single-phase transformer are marked H_1 and H_2. The leads on the low-voltage output or secondary side of each single-phase transformer are marked X_1 and X_2.

To connect the high-voltage primary windings in the closed-delta pattern to a three-phase source, the three windings are connected in series. In making the connection, the end of one primary winding is connected to the beginning of the next primary winding. In figure 24-3, H_1 is the beginning of each coil and H_2 is the end of each coil. Thus, each primary winding end H_2 is connected to the beginning H_1 of another primary winding. A three-phase line wire is also connected at each junction point H_1 – H_2. Note that the primary winding of each transformer is connected directly across the line voltage. This means that delta-connected transformers must be wound for the full line voltage. For figure 24-3, each of the three line voltages is 2,400 volts and the primary winding of each transformer is also rated at 2,400 volts. After the high-voltage primary connections are made, the three-phase, 2,400-volt input may be energized. It is not necessary to make polarity tests on the input side.

The next step is to connect the low-voltage output or secondary windings in the closed-delta pattern. The secondary winding leads are marked X_1 for the beginning of each coil and X_2 for the end of each coil. In making the connections on the secondary, the following procedure must be followed:

1. Check to see that the voltage output of each of the three transformers is 240 volts.

2. Connect the end of one secondary winding with the beginning of another secondary winding (figure 24-4).

The voltage across the open ends shown in figure 24-4 should be the same as the output of each transformer or 240 volts. If one of the transformers has its secondary winding connections reversed, the voltage across the open ends will be 1.73 × 240 = 415 volts.

Figure 24-5 illustrates an incorrect connection which must be changed so that it is the same as the connection shown in figure 24-4.

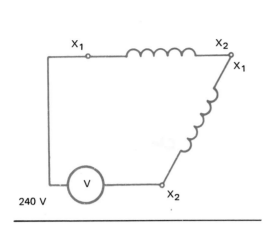

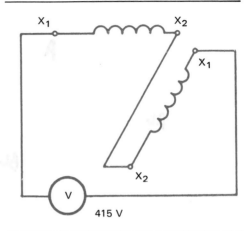

Fig. 24-4 A voltmeter is used to check correct connections.

Fig. 24-5 A voltmeter reading indicates incorrect connections.

Figure 24-6 illustrates the correct connections for the secondary coil of the third transformer. The voltage across the last two open ends should be zero if all the transformers are connected as shown. If the voltage is zero across the last two open ends, they may be connected together. A line lead is then connected at each of the three junction points $X_1 - X_2$. These three wires are the 240-volt, three-phase output. Note that each of the three line voltages and each of the three transformer output voltages is equal to 240 volts.

When the secondary winding of the third transformer is reversed, the voltage across the last two open leads is 240 + 240 = 480 volts.

Figure 24-7 illustrates the incorrect connection which results in a reading of 480 volts. The connections on the third transformer secondary must be reversed.

Caution: Never complete the last connection if there is a voltage difference greater than zero. If the connections are correct, this potential difference is zero. Observe safety precautions. De-energize the primary while making connections.

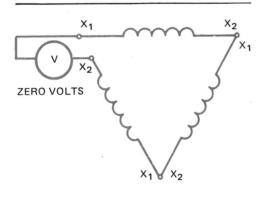

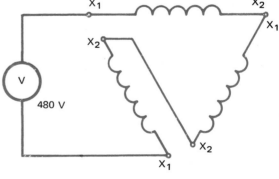

Fig. 24-6 Voltmeter reading indicates correct connections.

Fig. 24-7 Voltmeter reading indicates reversal of a coil.

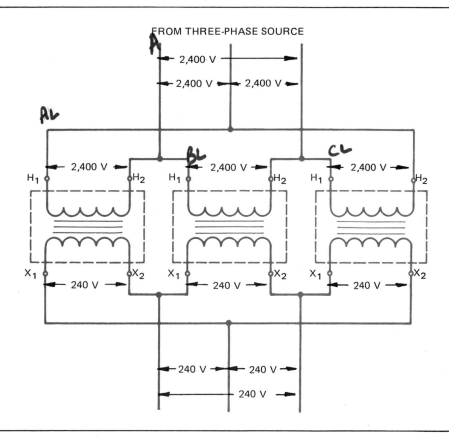

Fig. 24-8 Wiring diagram of delta-delta connection

When three transformers are connected with their primary windings in delta and their secondary windings in delta, the total connection is called a *delta-delta* ($\Delta - \Delta$) *connection.* The first delta symbol indicates the connection method of the primary windings, and the second delta symbol shows how the secondary windings are connected. When two or three single-phase transformers are used to step down or step up voltage on a three-phase system, the group is called a *transformer bank.*

Figure 24-8 is another way of showing the closed-delta connection first illustrated in figure 24-3. By tracing through the connection, it can be seen that the high-voltage and low-voltage windings are all connected in the closed-delta pattern. This type of transformer diagram is often used by the electrician.

VOLTAGE AND CURRENT

In any closed-delta transformer connection, two important facts must be kept in mind.

1. The line voltage and the voltage across the transformer windings are the same. A study of any delta connection shows that each transformer coil is connected directly across two line leads; therefore, the line voltage and the transformer coil voltage must be the same.

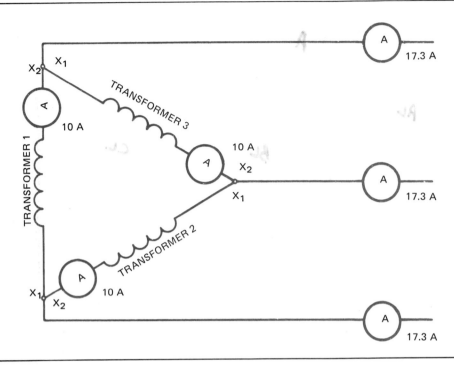

Fig. 24-9 Line current is $\sqrt{3}$ times the coil current in a delta connection.

2. The line current is greater than the coil current in a delta-connected transformer bank. The line current is equal to 1.73 × coil current. A study of a closed-delta transformer connection shows that each line lead is fed by two transformer coil currents which are out of phase and thus cannot be added directly.

In the arrangement shown in figure 24-9, the coil current in each transformer secondary is 10 amperes. The line current, however, is 1.73 × 10 or 17.3 amperes. Since the coil currents are out of phase, the total current is not 10 + 10 or 20 amperes. Rather, the total current is a resultant current in a balanced closed-delta system and is equal to 1.73 × coil current (1.73 equals the square root of three).

Three single-phase transformers of the same kilovolt-ampere (kVA) capacity are used in almost all delta-delta-connected transformer banks used to supply balanced three-phase industrial loads. For example, if the industrial load consists of three-phase motors, the current in each line wire is balanced. To determine the total kVA capacity of the entire delta-delta-connected transformer bank, add the three transformer kVA ratings. Thus, if each transformer is rated at 50 kVA, the total kVA is 50 + 50 + 50 = 150 kVA.

POWER AND LIGHTING SERVICE FROM A DELTA-DELTA-CONNECTED TRANSFORMER BANK

A delta-delta-connected transformer bank, with one transformer secondary center tapped, may be used to feed two types of load: (1) 240-volt, three-phase industrial power load, and (2) 120/240-volt, single-phase, three-wire lighting load.

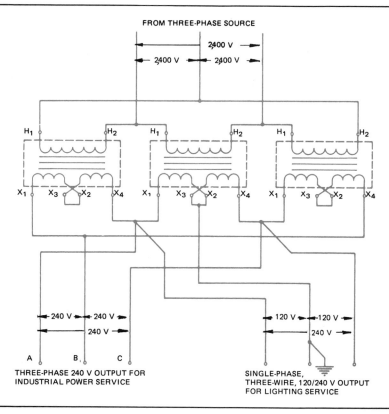

Fig. 24-10 Closed-delta transformer bank feeding a single-phase, three-wire lighting load and a three-phase, three-wire power load

The single-phase transformer which is to supply the single-phase, three-wire lighting load is usually larger in size than the other two transformers in the bank. This takes care of the additional lighting load placed here. A tap must be brought out from the midpoint of the 240-volt, low-voltage winding so that the 120/240-volt, single-phase, three wire service can be obtained. Many transformers are designed with the low-voltage side consisting of two 120-volt windings. These windings can be connected in series for 240 volts, in parallel for 120 volts, or in series with a tap brought out to give 120/240-volt service.

Figure 24-10 illustrates three single-phase transformers connected as a delta-delta transformer bank. Each transformer has two 120-volt, low-voltage windings. These 120-volt windings are connected in series to give a total output voltage of 240 volts for each transformer. The connection scheme for the high-voltage input or primary windings is closed delta. The low-voltage output or secondary windings are also connected in the closed-delta pattern to give three-phase, 240-volt service for the industrial power load. Note in figure 24-10 that the middle transformer is feeding the single-phase, three-wire, 120/240-volt lighting load. This center transformer has a mid tap on the secondary (output) side to give 120/240-volt service. Also note that this tap feeds to the grounded neutral wire.

The three-phase, 240-volt industrial power system is also connected to the transformer bank shown in figure 24-10. A check of the connections shows that both lines A and C of the three-phase, 240-volt system have 120 volts to ground. Line B, however, has 208 volts to ground (120 × 1.73 = 208). This situation is called the *high phase*.

Caution: The high-phase situation can be a serious hazard to human life as well as to any 120-volt equipment connected improperly between the high phase and neutral. When the voltage to ground exceeds 250 volts on any conductor in any metal raceway or metallic-sheathed cable, the National Electrical Code requires special bonding protection.

For example, if rigid conduit is used to connect the services, there must be two locknuts. One locknut is used outside and one inside any outlet box or cabinet. The regular conduit end bushing must also be used to protect the insulation on the wires in the conduit. Where the conductors are above a given size, this conduit bushing must be the insulating type or equivalent, according to the National Electrical Code in the section on cabinets.

Note that for ungrounded circuits, the greatest voltage between the given conductor and any other conductor of the circuit is considered the voltage to ground.

OPEN-DELTA OR V CONNECTION

A three-phase transformation of energy is possible using only two transformers. This connection arrangement is called the open-delta or V connection. The open-delta connection is often used in an emergency when one of the three transformers in a delta-delta bank becomes defective. When it is imperative that a consumer's three-phase power supply be restored as soon as possible, the defective transformer can be cut out of service using the open-delta arrangement.

The following example shows how the open-delta connection can be used in an emergency. Three 50-kVA transformers, each rated at 2,400 volts on the high-voltage winding and 240 volts on the low-voltage winding, are connected in a delta-delta bank. This closed-delta bank is used to step down a 2,400-volt, three-phase input to a 240-volt, three-phase output to supply an industrial consumer. Suddenly, the three-phase power service is interrupted because lightning strikes and damages one of the transformers. The service must be restored immediately. This situation is shown in figure 24-12.

Fig. 24-11 Three single-phase transformers, pole mounted. The three-phase secondary is on the bottom.

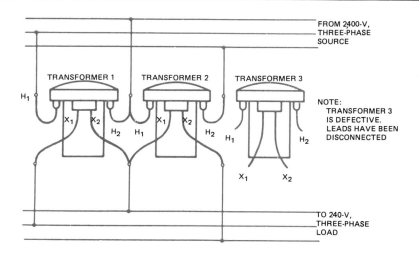

Fig. 24-12 Open-delta connection

Transformer 3 is the defective transformer. If all of the leads of the damaged transformer are disconnected, the closed-delta bank automatically becomes an open-delta transformer bank.

The schematic diagram of this open-delta connection is shown in figure 24-13. Note that with the one transformer removed, the triangular coil arrangement is open on one side. Because the schematic diagram resembles the letter V, this arrangement is also called the V connection.

While it appears that the total kVA of the open-delta bank should be two-thirds that of a closed-delta bank, the actual kVA rating of an open-delta bank is only 58 percent of the capacity of a closed delta bank. The reason for this is that the currents of the two transformers in the open delta connection are out of phase, resulting in the

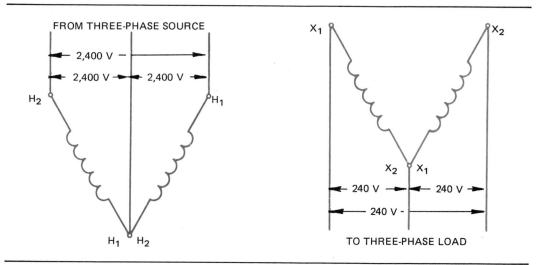

Fig. 24-13 Schematic diagram of the open-delta or V connection

total available capacity of the open-delta bank being only 58 percent instead of 66.7 percent.

In the open-delta example, there are three 50-kVA transformers connected in a delta-delta bank. This gives a total kVA capacity of 50 + 50 + 50 = 150 kVA for the closed-delta bank. When one transformer is disconnected, the transformer bank changes to an open-delta configuration, and the total kVA capacity now is only 58 percent of the original closed-delta capacity.

$$150 \times 0.58 = 87 \text{ kVA}$$

In some situations, an open-delta bank of transformers is installed initially. The third transformer is added when the increase in industrial power load on the transformer bank warrants the addition. When the third transformer is added to the bank, a closed-delta bank is formed.

When two transformers are installed in an open-delta configuration, the total bank capacity can be found by the use of the following procedure.

1. Add the two individual transformer kVA ratings. (For the problem given, the single-phase transformers are rated at 50 kVA.)

$$50 + 50 = 100 \text{ kVA}$$

2. Then multiply the total kVA value by 86.5 percent. This will give the total kVA capacity of the open-delta transformer bank.

$$100 \times 86.5\% = 87 \text{ kVA}$$

Therefore, an open-delta bank has a kVA capacity of 58 percent of the capacity of a closed-delta bank; an open-delta bank has a kVA capacity of 86.5 percent of the capacity of two transformers.

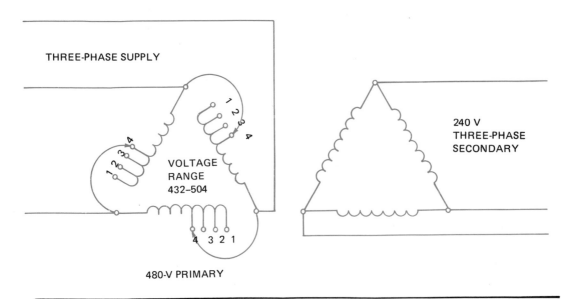

Fig. 24-14 Tap connections for a three-phase bank

THREE-PHASE TRANSFORMERS WITH PRIMARY TAPS

Some plant distribution transformers are pre-assembled and wired at the factory into a three-phase bank in a single enclosure or as a single unit. These assemblies consist of three single-phase transformers in one enclosure, usually the dry, air-cooled type. Some have primary tap terminals so that the supply voltage can be matched more closely (figure 24-14). The electrician must make the adjustment on the job until the primary of the transformer matches the measured supply voltage. The secondary will then produce the desired voltage to achieve a closer match of the equipment nameplate voltages. Utilities do not always supply the desired accurate voltages. There may also be a voltage drop within the plant.

When using taps on a three-phase transformer, or bank of transformers, it is important that the same taps on each of the three primaries be connected in the same position on each coil. (See Transformer Primary Taps in unit 22.) The following problems may result if the taps are not connected properly:

1. The output voltage on each of the three secondary voltages will not be the same. This will produce high unbalanced currents that will cause overheating of induction motors.

2. An undesirable circulating current will create a "false load" condition if the transformer is connected delta-delta.

Taps are used for consistently high or low voltages. They are not used with voltages that fluctuate or vary frequently.

ACHIEVEMENT REVIEW

1. What is one practical application of single-phase transformers connected in a delta-delta configuration? _____

_____

_____

2. What simple rule must be followed in making a delta connection? _____

_____

_____

_____

_____

_____

_____

3. Show a connection diagram for three single-phase transformers connected in a closed-delta scheme. This transformer bank is used to step down 2,400 volts, three-phase, to 240 volts, three-phase. Each transformer is rated at 50 kVA, with 2,400 volts on the high-voltage winding and 240 volts on the low-voltage winding. Mark leads H_1, X_1, and so forth. Show all voltages.

4. What is the total kVA capacity of the closed-delta transformer bank in question 3?

_____

5. What is one practical application of an open-delta transformer bank?

_____

_____

_____

_____

_____

6. Make a connection diagram of two single-phase transformers connected in open delta. Each transformer is rated at 10 kVA, with 4,800 volts on the high-voltage winding, and 240 volts on the low-voltage winding. This bank of transformers is to step down 4,800 volts, three phase, to 240 volts, three phase. Mark leads H_1, X_1, and so forth. Show all voltages. Calculate the total kVA capacity of this open-delta transformer bank.

7. What problems are likely to result if taps are not connected properly on a three-phase transformer bank? _____

_____

_____

_____

_____

SINGLE-PHASE TRANSFORMERS IN A WYE INSTALLATION

OBJECTIVES

After studying this unit, the student will be able to

- diagram the simple wye connection of three transformers.
- list the steps in the procedure for the proper connection and checking of the primary and secondary windings of three single-phase transformers connected in a wye arrangement.
- state the voltage and current relationships for wye-connected, single-phase transformers.
- describe how the grounded neutral of a three-phase, four-wire, wye-connected transformer bank maintains a balanced voltage across the windings.
- state how the kVA capacity of a wye-wye-connected transformer bank is obtained.

Voltage transformation on three-phase systems can also be accomplished using wye-connected single-phase transformers (figure 25-1). To avoid errors when wye connecting single-phase transformers, a systematic method of making the connections should be used. The electrician should know the basic voltage and current relationships common to this type of connection.

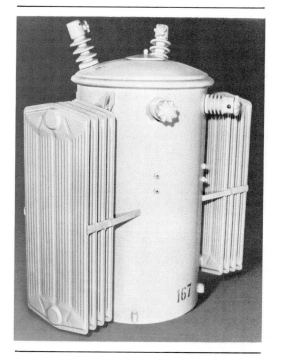

Fig. 25-1 Single-phase, round coil transformer *(Courtesy of McGraw-Edison Company, Power Systems Division)*

FUNDAMENTAL WYE CONNECTION

A simple wye system is formed by arranging three single-phase coils so that one end of each coil is connected at a common point (figure 25-2). Note that when these connections are shown in a schematic diagram, they resemble the letter Y (written wye). This configuration is also known as a "star" connection.

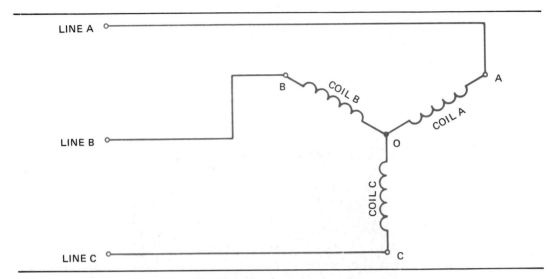

Fig. 25-2 Simple wye connection

As an example, figure 25-3 shows the wye-wye connection of three single-phase transformers to step down a three-phase input of 4,152 volts to a three-phase output of 208 volts. Each transformer must be voltage rated for its applications. The H_2 leads or primary winding ends of the transformer are connected together. The beginning or H_1 lead of each transformer is connected to one of the three line leads.

Two of the primary windings are in series across each pair of line wires. Each transformer primary winding is rated at 2,400 volts and the actual voltage applied to each of these three windings is 2,400 volts. Note that the potential across each pair of line leads is 4,152 volts and not the 2,400 + 2,400 = 4,800 volts which might be expected because of the series connection of two coils.

The value of 4,152 volts arises from the fact that the voltage applied to each of the primary windings is out of phase with the voltages applied to the other primary

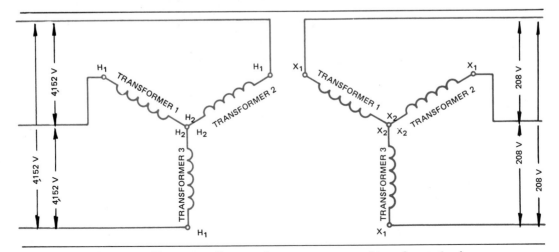

Fig. 25-3 Elementary diagram of wye-wye-connected transformer bank

windings. As a result, these winding voltages cannot be added directly to obtain the line voltage. Rather, the line voltage is equal to 1.73 × coil voltage. Therefore, for the input side of the transformer bank in figure 25-3 where the voltage on the primary winding of each transformer is 2,400 volts, the line voltage is

$$\begin{aligned} \text{Line voltage} \ &= \ 1.73 \times \text{coil voltage} \\ &= \ 1.73 \times 2,400 \\ &= \ 4,152 \text{ volts} \end{aligned}$$

If the coil voltage must be checked and the line voltage is known, the same value of 1.73 can be used. For this situation, the coil voltage is obtained by dividing the line voltage by 1.73.

$$\begin{aligned} \text{Coil voltage} \ &= \ \frac{\text{line voltage}}{1.73} \\ &= \ \frac{4,152}{1.73} \\ &= \ 2,400 \text{ volts} \end{aligned}$$

Thus, wye-connected transformer banks have only 58 percent of the line voltage applied to each of the three transformer windings. After the high-voltage primary connections are completed, the three-phase, 4,152-volt input may be energized. It is not necessary to make any polarity tests on the input (primary) side.

POLARITY TEST FOR UNMARKED AND NEW TRANSFORMERS

The next step is to connect the low-voltage output (secondary) windings in wye (figure 25-3). The following procedure must be followed when making the secondary connections.

1. Check to see that the voltage output of each of the three transformers is 120 volts.

    Caution: De-energize all circuits before making connections.

2. Connect the X_2 ends of two low-voltage secondary windings.

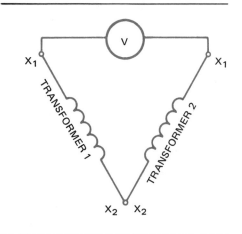

Fig. 25-4 Two transformers correctly connected

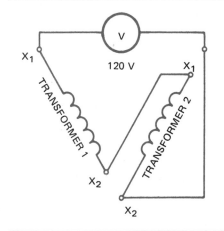

Fig. 25-5 Two transformers incorrectly connected

Figure 25-4 illustrates two secondary coils with the X_2 coil ends connected. The voltage across the open ends shoiud be 1.73 X 120 = 208 volts. However, if the leads on one transformer are reversed, the voltage across the open ends will be 120 volts.

Figure 25-5 illustrates two transformers connected incorrectly. The voltage across the open ends is only 120 volts. If the leads of transformer 2 are reversed, the connections will be correct and the voltage across the open ends will be 208 volts.

3. Connect the X_2 lead of the low-voltage secondary winding of transformer 3 with the X_2 leads of the other two transformers.

    The proper wye connection of the low-voltage secondary windings of the three single-phase transformers is shown in figure 25-6. The voltage across each pair of open ends should be 1.73 X 120 = 208 volts. If the voltage across the open ends is correct, then the line leads feeding to the three-phase, 208-volt secondary system may be connected.

Figure 25-7 illustrates the secondary windings connected in wye with the line leads properly connected. Since each of the line wires is connected in series with one of the transformer windings, the current in each winding is equal to its respective line current.

Whenever single-phase transformers are connected in wye, the following current and voltage relationships are true.

1. The line voltage is equal to 1.73 X winding voltage.

2. The line current and the winding current are equal.

The wye-wye connection scheme is satisfactory as long as the load on the secondary side is balanced. For example, this type of connection can be used if the load consists only of a three-phase motor load where the load currents are balanced. The wye-wye connection is unsatisfactory where the secondary load becomes greatly unbalanced.

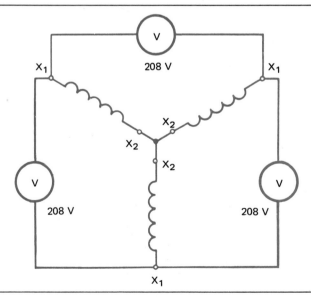

Fig. 25-6 Three single-phase transformers properly connected in a wye arrangement

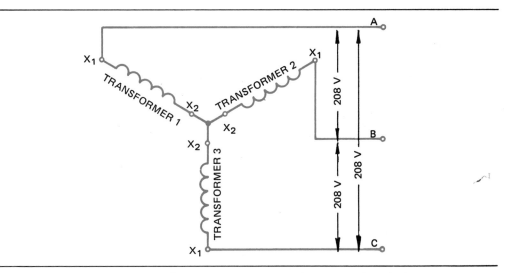

Fig. 25-7 Three single-phase transformers properly connected to the line

An unbalanced load results in a serious unbalance in the three output voltages of the transformer bank.

THREE-PHASE, FOUR-WIRE WYE CONNECTION

Voltage unbalancing in the secondary of the transformer bank can be nearly eliminated if a fourth wire (neutral wire) is used. This neutral wire connects between the source and the neutral point on the primary side of the transformer bank.

In the connection diagram (figure 25-8) a three phase, four-wire system is used to feed the three-phase, high-voltage input to the transformer bank. The grounded neutral

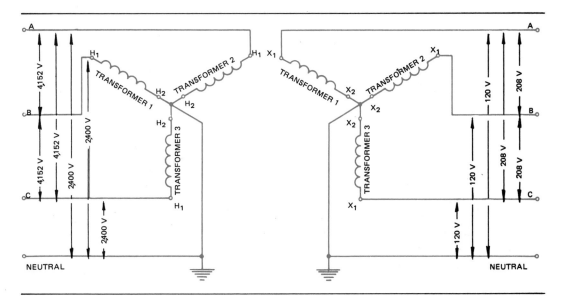

Fig. 25-8 Wye-wye transformer bank with neutral connection

Not used very often

wire is connected to the common point where all three high-voltage primary winding ends or H_2 leads connect. The voltage from the neutral to any one of the three line wires is 2,400 volts. Each high-voltage winding is connected between the neutral and one of the three line loads. Therefore, 2,400 volts is applied to each of the three high-voltage primary windings. The voltage across the three line leads is 1.73 × 2,400 volts or 4,152 volts. The neutral wire maintains a relatively constant voltage across each of the high-voltage primary windings even though the load is unbalanced. Because the neutral wire is grounded, it protects the three high-voltage primary windings from lightning surges.

A three-phase, four-wire system also feeds from the low-voltage secondary side of the transformer bank to the load. Each low-voltage secondary winding is connected between the secondary grounded neutral and one of the three line leads. As on the primary side, the grounded neutral protects the low-voltage secondary windings from lightning surges.

The voltage output of each secondary winding is 120 volts. The voltage between the neutral and any one of the three line leads on the secondary side is 120 volts, as shown in figure 25-8. The voltage across the three line leads is 1.73 × 120 = 208 volts. Thus, by using a three-phase, four-wire secondary, two voltages are available for different types of loads: 208 volts, three phase, for industrial power loads such as three-phase motors; and 120 volts, three phase, for lighting loads.

Many single-phase transformers are designed so that the low-voltage side consists of two 120-volt windings. These two windings can be connected in series for 240 volts or in parallel for 120 volts.

Figure 25-9 shows three single-phase transformers connected as a wye-wye bank. Each transformer has two 120-volt, low-voltage windings. For each single-phase transformer, the low-voltage coils are connected in parallel to give a voltage output of 120 volts. Note in figure 25-9 that the secondary output windings of the three transformers are connected in wye. This three-phase, four-wire secondary system provides two different types of service:

- three-phase, 208-volt service for motor loads

- single-phase, 120-volt service for lighting loads

The 120/208-volt wye system is commonly used in schools, stores, and offices. Another popular system for large installations is the 480/277-volt wye system. Some applications of this system include:

- motors connected to 480 volts (phase to phase);

- certain fluorescent fixtures connected to 277 volts (phase to neutral);

- 120-volt outlets, incandescent lamps, and appliances connected to 120-volt circuits supplied from single-phase, 480/120/240-volt transformers or three-phase, 480/208Y/120-volt transformers. These separate transformers are connected to the 480-volt feeders for the primary source.

Three single-phase transformers of the same kilovolt-ampere capacity are used in most wye-wye-connected transformer banks. The total kilovolt-ampere capacity of a wye-wye-connected bank is found by adding the individual kVA ratings of the transformer. If each transformer is rated at 25 kVA, then the total kVA is 25 + 25 + 25 = 75 kVA.

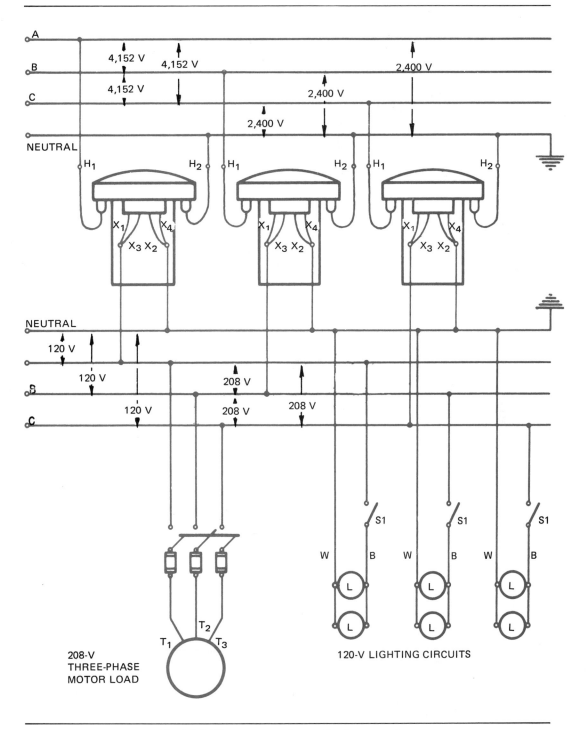

Fig. 25-9 Wye-wye transformer bank connections

If one transformer becomes defective, it must be replaced before the transformer bank can be reenergized. A wye-wye-connected transformer bank cannot be reconnected in an emergency situation using only two single-phase transformers, such as in the open-delta system.

ACHIEVEMENT REVIEW

1. Draw a connection diagram for three wye-wye-connected single-phase transformers. This transformer bank is used to step down 2,400/4,152 volts on a three-phase, four-wire primary to 120/208 volts on a three-phase, four-wire secondary. Each transformer is rated at 20 kVA, with 2,400 volts on the high-voltage winding and 120 volts on the low-voltage winding. Mark leads H_1, X_1, and so forth; show all voltages.

2. What is the total kVA capacity of the wye-wye transformer bank in question 1?

    _____

3. A grounded neutral wire is used with a wye-wye-connected transformer bank for what purpose? _____

    _____

4. The three-phase, four-wire secondary output of a wye-connected transformer bank can be used for what two types of load?

    a. _____

    _____

    b. _____

    _____

5. List the steps that may be used in connecting three single-phase transformers in wye.

a. _____

_____

_____

b. _____

_____

_____

c. _____

_____

_____

_____

6. When single-phase transformers are connected in a three-phase Y,

a. what is the line current compared with the phase-winding current?

_____

b. what is the line voltage compared with the phase-winding voltage?

_____

WYE AND DELTA CONNECTIONS OF SINGLE-PHASE TRANSFORMERS

OBJECTIVES

After studying this unit, the student will be able to

- diagram the connection of three single-phase transformers to form a delta-wye transformer bank.

- describe how a delta-wye transformer bank is used to step down voltages.

- describe how a delta-wye transformer bank is used to step up voltages.

- diagram the connection of three single-phase transformers to form a wye-delta transformer bank.

- describe how a wye-delta transformer bank is used to step down voltages.

- list advantages and disadvantages of a single three-phase transformer as compared to three single-phase transformers.

Five commonly-used methods of connecting single-phase transformers to form three-phase transformer banks are: delta-delta, open delta, wye-wye, delta-wye, and wye-delta connections. The delta-delta, open delta, and wye-wye connections are described in previous units.

This unit covers the delta-wye and wye-delta connections. The current and voltage relationships for each of these methods of three-phase transformation are explained, and examples of several applications for each connection method are shown.

STEP-DOWN APPLICATION FOR DELTA-WYE TRANSFORMER BANK

Assume that electrical energy must be transformed from a 2,400-volt, three-phase, three-wire input to a 120/208-volt, three-phase, four-wire output. Each of the three single-phase transformers is rated at 20 kVA, with 2,400 volts on the high-voltage windings and 120 or 240 volts on the low-voltage windings.

The primary windings of the three single-phase transformers are delta connected. The line voltage of the three-phase, three-wire primary input is 2,400 volts. Remember that the line voltage and the coil voltage are the same in a delta connection. As a result, the voltage across each of the primary coil windings is also 2,400 volts.

Figure 26-1 illustrates the connections for the delta-wye transformer bank in this example. Each transformer has two 120-volt, low-voltage windings which are connected in parallel to give a voltage output of 120 volts for each single-phase transformer. The secondary connections show that the output windings of the three transformers are connected in wye. Two types of service are available as a result of the three-phase, four-wire secondary system:

- three-phase, 208-volt service for motor loads;
- single-phase, 120-volt service for lighting loads.

The primary or input side of the bank in figure 26-1 is delta connected. Therefore, for the primary side of the delta bank, the following are true:

- the line voltage and the coil voltage are the same;
- the line current is equal to 1.73 × coil current.

The secondary or output side of this transformer bank is wye connected. For the secondary, then:

- the line voltage is equal to 1.73 × coil voltage;
- the line current and the coil current are equal.

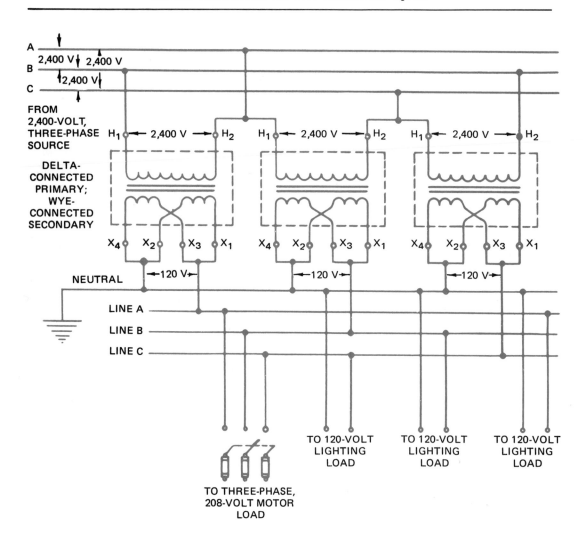

Fig. 26-1 Delta-wye transformer bank

The three single-phase transformers used in a delta-wye connection have the same kVA capacity. The transformers in this problem are each rated at 20 kVA. The total kVA capacity of a delta-wye transformer bank is determined by adding the three kVA ratings. Since each transformer is rated at 20 kVA, the total delta-wye bank capacity is 60 kVA.

If one transformer becomes defective, it must be replaced before the bank can be reenergized. In an emergency situation, a delta-wye-connected transformer bank cannot be reconnected using only two transformers.

In the delta-wye connection illustrated in figure 26-1, the three single-phase transformers are connected to obtain additive polarity. However, the transformers that are used in an installation may have either additive or subtractive polarity. The polarity of each transformer must be checked. Then, if the basic rules for making delta connections and wye connections are followed, the electrician should have no difficulty in making any standard three-phase transformer bank connections.

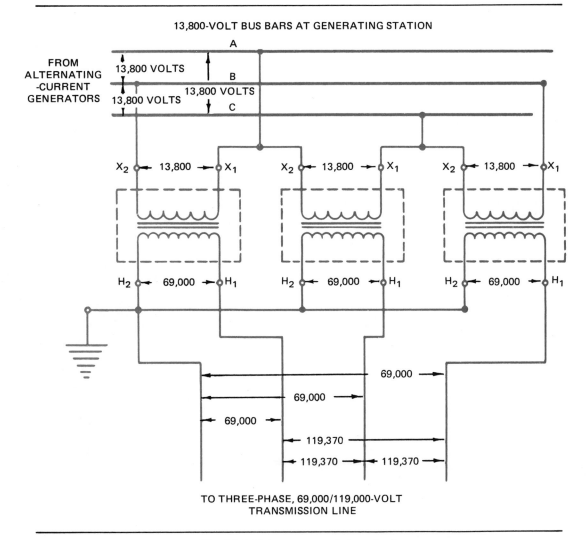

Fig. 26-2 Delta-wye transformer bank

STEP-UP APPLICATION FOR DELTA-WYE TRANSFORMER BANK

The delta-wye transformer bank is well adapted for stepping up voltages. The input voltage is stepped up by the transformer ratio and then is increased further by the voltage relationship for a wye connection: line voltage = 1.73 × coil voltages. In addition, the insulation requirements for the secondaries are reduced. This is an important advantage when very high voltages are used on the secondary side.

A delta-wye transformer bank used to step up the voltage at a generating station is illustrated in figure 26-2. The high voltage output is connected to three-phase transmission lines. These transmission lines deliver the electrical energy to municipal and industrial consumers who may be miles away from the generating station.

As shown in figure 26-2, the alternating-current generators deliver energy to the generating station bus bars at a three-phase potential of 13,800 volts. The primary windings of the three single-phase transformers are each rated at 13,800 volts. These primary windings are connected in delta to the generating station bus bars; therefore, each primary coil winding has 13,800 volts applied to it. The transformers have a step-up ratio of 1 to 5. As a result, the voltage output of the secondary of each single-phase transformer is 5 × 13,800 = 69,000 volts. Figure 26-2 shows that the three secondary

Fig. 26-3 Wye-delta transformer bank (Two 2,500-kVA, 1,200–2,400-V, wye-delta duplicate transformers)

windings are connected in wye. Each high-voltage secondary winding is connected between the secondary neutral and one of the three line leads. The voltage between the neutral and any one of the three line leads is the same as the secondary coil voltage or 69,000 volts. The voltage across the three line leads is 1.73 X 69,000 = 119,370 volts. The grounded neutral wire on the high-voltage secondary output must be used to obtain balanced three-phase voltages even though the load current may be unbalanced. Not only is this neutral wire grounded at the transformer bank, it is also grounded at periodic intervals on the transmission line. As a result, it protects the three high-voltage secondary windings of the single-phase transformers from possible damage due to lightning surges.

WYE-DELTA TRANSFORMER BANK

A transformer bank connected in wye-delta is the type most often used to step down relatively high transmission line voltages (60,900 volts or more) at the consumer's location. Two reasons for selecting this type of transformer bank are that the three-phase voltage is decreased by the transformer ratio multiplied by the factor 1.73, and the insulation requirements for the high-voltage primary windings are reduced.

As an example, assume that it is necessary to step down a three-phase 60,900-volt input to a three-phase, 4,400-volt output (figure 26-4). The primary windings

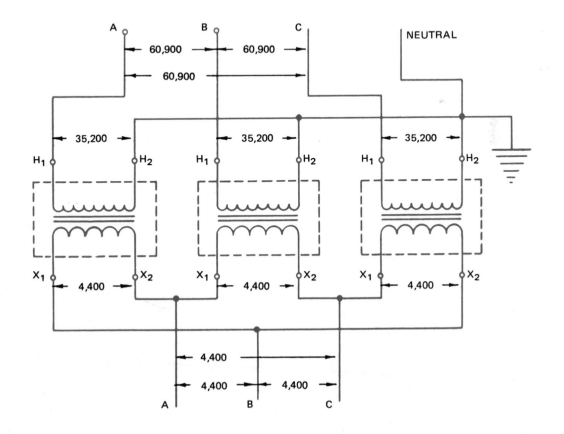

Fig. 26-4 Wye-delta transformer bank

are connected in wye to a three-phase, four-wire transmission line. The three line voltages are 60,900 volts each and the voltage from each line wire to the grounded neutral is 35,200 volts.

Each of the three single-phase transformers is rated at 1,000 kVA, with 35,200 volts on the high-voltage side and 4,400 volts on the low-voltage side. The voltage ratio of each transformer is 8 to 1.

Figure 26-4 shows that the secondary windings are connected in delta, resulting in a line voltage of 4,400 volts on the three-phase, three-wire secondary system feeding to the load.

The grounded neutral protects the three high-voltage primary windings of the transformers from possible damage due to lightning surges.

The total kVA capacity of a wye-delta transformer bank is determined by adding the kVA rating of each single-phase transformer in the bank. For the bank in figure 26-4, the total kVA capacity is equal to 1,000 + 1,000 + 1,000 = 3,000 kVA.

THREE-PHASE TRANSFORMERS

Voltages on three-phase systems may be transformed using three-phase transformers. The core of a three-phase transformer is made with three legs. A primary and a secondary winding of one phase are placed on each of the three legs. These transformers may be connected in delta-delta, wye-wye, delta-wye or wye-delta. The connections are made inside the transformer case. For delta-delta connections, three high-voltage and three

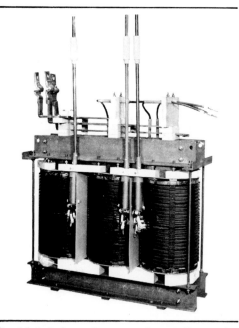

Fig. 26-5 A three-phase transformer (Assembled core and coils for a 500-kVA, 60-Hz, 13,800-to 2,400-V transformer)

Fig. 26-6   A 5,000-kVA, 34-kV power transformer *(Courtesy of R. E. Uptegraff Manufacturing Company)*

low-voltage leads are brought out. Four leads are brought out when any wye-connected windings are used. This fourth lead is necessary for the neutral wire connection.

The three-phase transformer occupies less space than three single-phase transformers because the windings can be placed on one core in the three-phase transformer case, (figure 26-5). The efficiency of a three-phase transformer is higher than the overall efficiency of three single-phase transformers connected in a three-phase bank.

However, there is one disadvantage to the use of a three-phase transformer. If one of the phase windings becomes defective, the entire three-phase unit must be taken out of service. If a single-phase transformer in a three-phase bank becomes defective it can be replaced quickly. The resultant power interruption is brief. For this reason, most transformer installations consist of banks of three single-phase transformers.

ACHIEVEMENT REVIEW

1. Diagram the connections for three single-phase transformers connected in delta-wye to step down 2,400 volts, three phase, three wire, to a 120/208-volt, three-phase, four-wire service. Three single-phase transformers are to be used. Each transformer is rated at 25 kVA, with 2,400 volts on the high-voltage side and 120 volts on the low-voltage side. Mark leads H_1, X_1, and so forth. Show all voltages.

2. What is the total kVA capacity of the delta-wye transformer bank in question 1? _____

3. What are two applications of a three-phase, delta-wye transformer bank?

    a. _____

    _____

    _____

    b. _____

    _____

    _____

4. What is one practical application of a three-phase, wye-delta transformer bank?

    _____

    _____

    _____

5. Diagram the connections for three single-phase transformers connected in wye-delta to step down a three-phase input of 33,000 volts to a three-phase output of 4,800 volts. Mark leads H_1, X_1, and so forth. Show all voltages.

6. What is one advantage to the use of a three-phase transformer in place of three single-phase transformers? _____

_____

_____

_____

7. What is one disadvantage to the use of a three-phase transformer in place of three single-phase transformers? _____

_____

_____

_____

8. Insert the word or phrase that completes each of the following statements.

   a. A wye-delta transformer bank has _____ -connected primary windings and _____ - connected secondary windings.

   b. A delta-wye transformer bank has _____ -connected primary windings and _____ - connected secondary windings.

   c. A wye-delta transformer bank is used to _____ extremely high three-phase voltages.

   d. A three-phase transformer takes _____ space than a transformer bank of the same kVA capacity consisting of three single-phase transformers.

   e. A three-phase transformer has a _____ percent efficiency than a transformer bank consisting of three single-phase transformers.

9. List five common three-phase connections used to connect transformer banks consisting of either two or three single-phase transformers.

   a. _____

   b. _____

   c. _____

   d. _____

   e. _____

10. What is the purpose of the grounded neutral on a three-phase, four-wire system?

_____

_____

_____

Select the correct answer for each of the following statements and place the corresponding letter in the space provided.

11. A delta-wye, four-wire secondary gives   _____
    a.   120-volt, single-phase and 208-volt, three-phase output.
    b.   208-volt, single-phase and 120-volt, three-phase output.
    c.   208-volt, three-phase output.
    d.   120-volt, three-phase output.

12. Most three-phase systems use three single-phase transformers
    connected in a bank because   _____
    a.   one transformer can be replaced readily if it becomes
         defective.
    b.   better regulation is maintained.
    c.   they are easier to cool.
    d.   this method of connection is the most efficient.

13. Transformer capacities may be increased by   _____
    a.   connecting them in series.
    b.   pumping the oil.
    c.   cooling the oil with fans.
    d.   reducing the load.

14. A step-down, delta-wye transformer connection is commonly
    used for   _____
    a.   motor and lighting loads.
    b.   distribution of electrical energy.
    c.   motor loads.
    d.   lighting loads.

15. In a delta connection   _____
    a.   the line voltage and coil voltage are equal.
    b.   the line current is equal to 1.73 times the coil current.
    c.   the coils are connected in closed series.
    d.   all of these are true.

16. In a wye connection   _____
    a.   the line current is equal to 1.73 times the coil current.
    b.   the line voltage is equal to 1.73 times the coil voltage.
    c.   the line voltage and coil voltage are equal.
    d.   none of these is true.

INSTRUMENT TRANSFORMERS

OBJECTIVES

After studying this unit, the student will be able to

- explain the operation of an instrument potential transformer.
- explain the operation of an instrument current transformer.
- diagram the connections for a potential transformer and a current transformer in a single-phase circuit.
- state how the following quantities are determined for a single-phase circuit containing instrument transformers: primary current, primary voltage, primary power, apparent power, and power factor.
- describe the connection of instrument transformers in a three-phase, three-wire circuit.
- describe the connection of instrument transformers to a three-phase, four-wire system.

Instrument transformers are used in the measurement and control of alternating-current circuits. Direct measurement of high voltage or heavy currents involves large and expensive instruments, relays, and other circuit components of many designs. The use of instrument transformers, however, makes it possible to use relatively small and inexpensive instruments and control devices of standardized designs. Instrument transformers also protect the operator, the measuring devices, and the control equipment from the dangers of high voltage. The use of instrument transformers results in increased safety, accuracy, and convenience.

There are two distinct classes of instrument transformers: the instrument potential transformer and the instrument current transformer. (The word "instrument" is usually omitted for brevity.)

POTENTIAL TRANSFORMERS

The potential transformer operates on the same principle as a power or distribution transformer. The main difference is that the capacity of a potential transformer is small compared to that of power transformers. Potential transformers have ratings from 100 to 500 volt-amperes (VA). The low-voltage side is usually wound for 115 volts. The load on the low-voltage side usually consists of the potential coils of various instruments, but may also include the potential coils of relays and other control equipment. In general, the load is relatively light and it is not necessary to have potential transformers with a capacity greater than 100 to 500 volt-amperes.

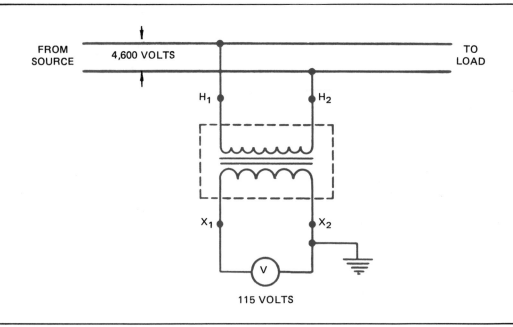

Fig. 27-1 Connections for a potential transformer

The high-voltage primary winding of a potential transformer has the same voltage rating as the primary circuit. Assume that it is necessary to measure the voltage of a 4,600-volt, single-phase line. The primary of the potential transformer is rated at 4,600 volts and the low-voltage secondary is rated at 115 volts. The ratio between the primary and secondary windings is:

$$\frac{4,600}{115} \quad \text{or} \quad \frac{40}{1}$$

A voltmeter connected across the secondary of the potential transformer indicates a value of 115 volts. To determine the actual voltage on the high-voltage circuit, the instrument reading of 115 volts must be multiplied by 40: 115 × 40 = 4,600 volts. In some cases, the voltmeter is calibrated to indicate the actual value of voltage on the primary side. As a result, the operator is not required to apply the multiplier to the instrument reading, and the possibility of errors is reduced.

Figure 27-1 illustrates the connections for a potential transformer with a 4,600-volt primary input and a 115-volt output to the voltmeter. This potential transformer has subtractive polarity. (All instrument potential transformers now manufactured have subtractive polarity.) One of the secondary leads of the transformer in figure 27-1 is grounded to eliminate high-voltage hazards.

Potential transformers have highly accurate ratios between the primary and secondary voltage values; generally the error is less than 0.5 percent.

CURRENT TRANSFORMERS

Current transformers are used so that ammeters and the current coils of other instruments and relays need not be connected directly to high-voltage lines. In other words, these instruments and relays are insulated from high voltages. Current transformers also step down the current in a known ratio. The use of current transformers means that relatively small and accurate instruments, relays, and control devices of standardized design can be used in circuits.

The current transformer has separate primary and secondary windings. The primary winding, which consists of a few turns of heavy wire wound on a laminated iron core, is connected in series with one of the line wires. The secondary winding consists of a greater number of turns of a smaller size of wire. The primary and secondary windings are wound on the same core.

The current rating of the primary winding of a current transformer is determined by the maximum value of the load current. The secondary winding is rated at 5 amperes regardless of the current rating of the primary windings.

Assume that the current rating of the primary winding of a current transformer is 100 amperes. The primary winding has three turns and the secondary winding has 60 turns. The secondary winding has the standard current rating of 5 amperes; therefore, the ratio between the primary and secondary currents is 100/5 or 20 to 1. The primary current is 20 times greater than the secondary current. Since the secondary winding has 60 turns and the primary winding has 3 turns, the secondary winding has 20 times as many turns as the primary winding. For a current transformer, then, the ratio of primary to secondary currents is inversely proportional to the ratio of primary to secondary turns.

In figure 27-2, a current transformer is used to step down current in a 4,600-volt, single-phase circuit. The current transformer is rated at 100 to 5 amperes and the ratio of current step down is 20 to 1. In other words, there are 20 amperes in the primary winding for each ampere in the secondary winding. If the ammeter at the secondary indicates 4 amperes, the actual current in the primary is 20 times this value or 80 amperes.

The current transformer in figure 27-2 has polarity markings in that the two high-voltage primary leads are marked H_1 and H_2, and the secondary leads are marked X_1 and X_2. When H_1 is instantaneously positive, X_1 is positive at the same moment. Some

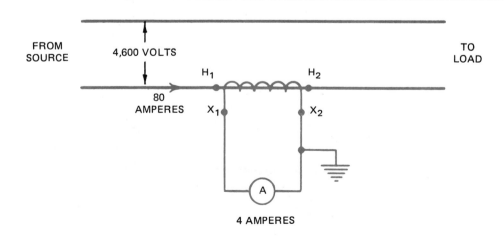

Fig. 27-2 A current transformer used with an ammeter

current transformer manufacturers mark only the H_1 and X_1 leads. When connecting current transformers in circuits, the H_1 lead is connected to the line lead feeding from the source, while the H_2 lead is connected to the line lead feeding to the load. The secondary leads are connected directly to the ammeter. Note that one of the secondary leads is grounded as a safety precaution to eliminate high-voltage hazards.

Caution: The secondary circuit of a transformer should never be opened when there is current in the primary winding. If the secondary circuit is opened when there is current in the primary winding, then the entire primary current is an exciting current which induces a high voltage in the secondary winding. This voltage can be high enough to endanger human life.

Individuals working with current transformers must check that the secondary winding circuit path is closed. At times, it may be necessary to open the secondary instrument circuit when there is current in the primary winding. For example, the metering circuit may require rewiring or other repairs may be needed. To protect the worker, a small short-circuiting switch is connected into the circuit at the secondary terminals of

Fig. 27-3 A portable current transformer *(Courtesy of A & M Instruments)*

the current transformer. This switch is closed when the instrument circuit must be opened for repairs or rewiring.

Current transformers have very accurate ratios between the primary and secondary current values: the error of most modern current transformers is less than 0.5 percent.

When the primary winding has a large current rating it may consist of a straight conductor passing through the center of a hollow metal core. The secondary winding is wound on the core. This assembly is called a bar-type current transformer. The name is derived from the construction of the primary which actually is a straight copper bus bar. All standard current transformers with ratings of 1,000 amperes or more are bar-type transformers. Some current transformers with lower ratings may also be of the bar type. Figure 27-3 shows a portable current transformer.

Fig. 27-4 Clamp-on current ammeter and voltmeter *(Courtesy of A & M Instruments)*

INSTRUMENT TRANSFORMERS IN A SINGLE-PHASE CIRCUIT

Figure 27-5 illustrates an instrument load connected through instrument transformers to a single-phase, high-voltage line. The instruments include a voltmeter (figure 27-6), an ammeter, and a wattmeter. The potential transformer is rated at 4,600 to 115 volts; the current transformer is rated at 50 to 5 amperes. The potential coils of the voltmeter and the wattmeter are connected in parallel across the low-voltage output of the potential transformer. Therefore, the voltage across the potential coils of each of these instruments is the same. The current coils of the ammeter and the wattmeter are connected in series across the secondary output of the current transformer. As a result, the current in the current coils of both instruments is the same. Note that the secondary of each instrument transformer is grounded to provide protection from high-voltage hazards.

The voltmeter in figure 27-5 reads 112.5 volts, the ammeter reads 4 amperes, and the wattmeter reads 450 watts. To find the primary voltage, primary current, primary power, apparent power in the primary circuit and the power factor, the following procedures are used:

Primary Voltage

$$\text{Voltmeter multiplier} = 4{,}600/115 = 40$$
$$\text{Primary volts} = 112.5 \times 40$$
$$= 4{,}500 \text{ volts}$$

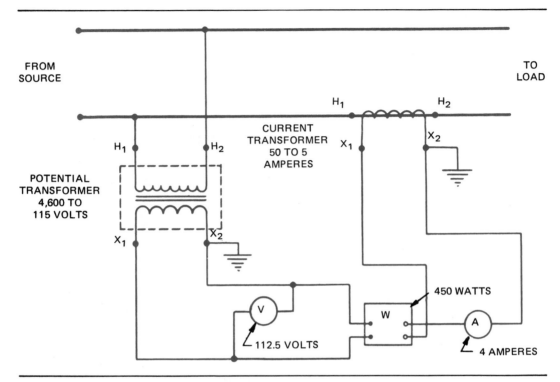

Fig. 27-5 Single-phase metering connections

Fig. 27-6 A switchboard-type voltmeter *(Courtesy of General Electric Company)*

Primary Current

Ammeter multiplier = 50/5 = 10
Primary amperes = 4 × 10
= 40 amperes

Primary Power

Wattmeter multiplier = Voltmeter multiplier × ammeter multiplier
Wattmeter multiplier = 40 × 10
= 400
Primary watts = 450 × 400
= 180,000 watts or 180 kilowatts

Apparent Power

The apparent power of the primary circuit is found by multiplying the primary voltage and current values.

Apparent power (volt-amperes) = volts × amperes
volt-amperes = 4,500 × 40
= 180,000 watts = $\frac{180,000}{1,000}$ = 180 kilowatts

Power Factor

Power factor = $\frac{\text{Power in kilowatts}}{\text{Apparent power in kilovolt-amperes}}$
= 180/180
= 1.00 or 100 percent

INSTRUMENT TRANSFORMERS ON THREE-PHASE SYSTEMS

Three-Phase, Three-Wire System

On a three-phase, three-wire system, two potential transformers of the same rating and two current transformers of the same rating are necessary. It is common practice in three-phase metering to interconnect the secondary circuits. That is, the connections are made so that one wire or device conducts the combined currents of two transformers in different phases.

The low-voltage instrument connections for a three-phase, three-wire system are shown in figure 27-7. Note that the two potential transformers are connected in open delta to the 4,600-volt, three-phase line. This results in three secondary voltage values of 115 volts each. The two current transformers are connected so that the primary of one transformer is in series with line A and the primary winding of the second transformer is in series with line C.

Note that three ammeters are used in the low-voltage secondary circuit. This wiring system is satisfactory on a three-phase, three-wire system and all three ammeters give accurate readings. Other instruments which can be used in this circuit include a three-phase wattmeter, a three-phase watthour meter, and a three-phase power factor

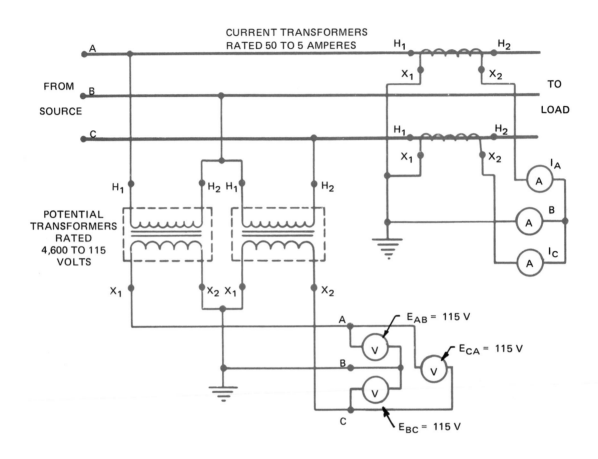

Fig. 27-7 Metering connections for three-phase, three-wire system

meter. When three-phase instruments are connected in the secondary circuits, these instruments must be connected correctly so that the proper phase relationships are maintained. If this precaution is not observed, the instrument readings will be incorrect. In checking the connections for this three-phase, three-wire metering system, note that the interconnected potential and current secondaries are both grounded to provide protection from high-voltage hazards.

Three-Phase, Four-Wire System

Figure 27-8 illustrates the secondary metering connections for a 2,400/4,152-volt, three-phase, four-wire system. The three potential transformers are connected in wye to give a three-phase output of three secondary voltages of 120 volts to neutral. Three 50-to-5-ampere current transformers are used in the three line conductors. Three ammeters are used in the interconnected secondary circuit. Both the interconnected potential and the current secondaries are grounded to protect against possible high-voltage hazards.

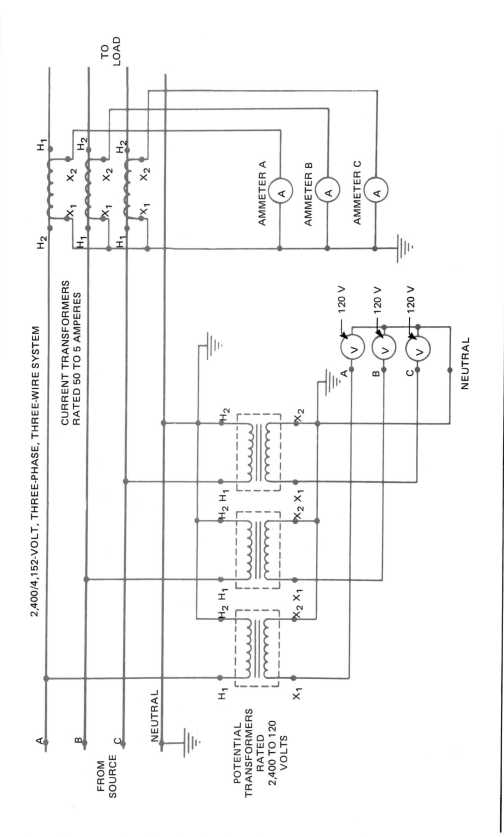

Fig. 27-8 Metering connections for three-phase, four-wire system

ACHIEVEMENT REVIEW

1. What are the two types of instrument transformers?

   a. _____

   b. _____

2. Why must the secondary circuit of a current transformer be closed when there is current in the primary circuit?

   _____

   _____

3. A transformer is rated at 4,600/115 volts. A voltmeter connected across the secondary reads 112 volts. What is the primary voltage? _____

4. A current transformer is rated at 150/5 amperes. An ammeter in the secondary circuit reads 3.5 amperes. What is the primary current? _____

5. A 2,300/115-volt potential transformer and a 100/5-ampere current transformer are connected on a single-phase line. A voltmeter, an ammeter, and a wattmeter are connected in the secondaries of the instrument transformers. The voltmeter reads 110 volts, the ammeter reads 4 amperes, and the wattmeter reads 352 watts. Draw the connections for this circuit. Mark leads H_1, X_1, and so forth. Show all voltage, current, and wattage readings.

6. Complete a circuit using instrument transformers to measure voltage and amperage. Include terminal markings.

FROM SOURCE   TO LOAD

7. What is the primary voltage of the single-phase circuit given in question 5?

_____

8. What is the primary current in amperes of the single-phase circuit given in question 5?

_____

9. What is the primary power in watts in the single-phase circuit given in question 5?

_____

10. What is the power factor of the single-phase circuit given in question 5?

_____

Select the correct answer for each of the following statements and place the corresponding letter in the space provided.

11. The secondary for a potential transformer is usually wound for

_____

    a. 10 volts.   c. 230 volts.
    b. 115 volts.   d. 500 volts.

12. Potential transformer secondaries are grounded to

_____

    a. stabilize meter readings.
    b. insure readings with an accuracy of 0.5 percent.
    c. complete a system with the primaries.
    d. eliminate high-voltage hazards.

13. A transformer used to reduce current values to a size where
    small meters can register them is a(n)   _____
    a. autotransformer.   c. potential transformer.
    b. distribution transformer.   d. current transformer.

14. The primary of a large current transformer may consist of   _____
    a. many turns of fine wire.
    b. few turns of fine wire.
    c. many turns of heavy wire.
    d. straight-through conductor.

15. The standard ampere rating of the secondary of a current
    transformer is   _____
    a. 5 amperes.   c. 15 amperes.
    b. 50 amperes.   d. 15 amperes.

16. The secondary circuit of a current transformer should never
    be opened when current is present in the primary because   _____
    a. the meter will burn out.
    b. the meter will not operate.
    c. dangerous high voltage may develop.
    d. primary values may be read on the meter.

NATIONAL ELECTRICAL CODE REQUIREMENTS FOR TRANSFORMER INSTALLATIONS

UNIT 28

OBJECTIVES

After studying this unit, the student will be able to

- use the National Electrical Code to determine the requirements and limitations of transformer installations.

The National Electrical Code (NEC) covers the minimum requirements of the installation of electrical wiring and equipment within public or private buildings and their premises.

TRANSFORMER LOCATION

The location of transformers is a prime Code ruling. Most electrical codes and power companies state that transformers and transformer vaults must be readily accessible to qualified personnel for inspection and maintenance. The codes also contain specific sections covering oil-insulated, askarel-insulated, other dielectric fluids, and dry-type transformers, as well as transformer vaults. Dry-type transformers installed outdoors should have weather-proof enclosures.

TRANSFORMER OVERCURRENT PROTECTION

The National Electrical Code gives information on the overcurrent protection required for transformers and transformer banks, as well as the maximum overcurrent protection allowed on the primary of a transformer.

Figure 28-1 illustrates where transformer primary protection is located.

For a transformer 600 volts or less, overcurrent protection is permitted in the secondary in place of primary protection provided that certain regulations are followed.

1. The secondary overcurrent protection must not be greater than 125 percent of the rated current of the secondary.

2. The primary feeder must not have overcurrent protection in excess of six times the rated primary current of the transformer. This is allowable only when the percentage impedance of the transformer is not in excess of 6 percent.

3. If the percentage impedance is greater than 6 percent but less than 10 percent, the primary feeder must not be rated in excess of four times the rated primary current of the transformer.

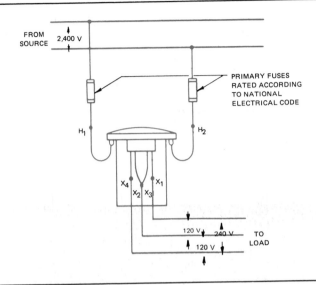

Fig. 28-1 Transformer overcurrent protection

About 90 percent of the transformers used within buildings are the dry type and generally require lower values of overcurrent protection. Consult *Section 450-3* of the National Electrical Code for specific information as to the value of the protection.

Figure 28-2 illustrates a transformer connection where the overcurrent protection is inserted in the secondary circuit. Note that the only primary protection is the feeder overcurrent protection.

A transformer with integral thermal overload protection (the protection is built into the transformer) does not need primary fuse protection. However, there must be

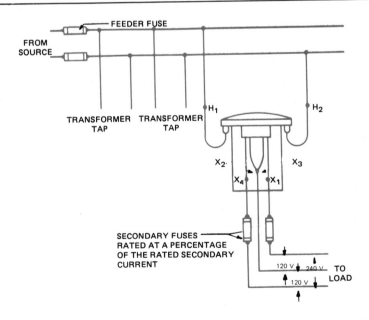

Fig. 28-2 Transformer and feeder overcurrent protection

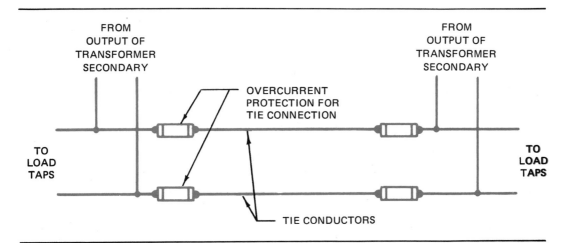

Fig. 28-3 Tie connections between transformers

primary feeder protection. The feeder overcurrent rating requirements are the same as given previously.

The Code requires that instrument potential transformers have primary fuses. Fuses of different sizes are required for operation above and below 600 volts.

SECONDARY CONNECTIONS BETWEEN TRANSFORMERS

The Code defines a secondary tie as a circuit operating at 600 volts or less between phases. This circuit connects two power sources or power supply points such as the secondaries of two transformers.

A secondary tie circuit should have overcurrent protection at each end except in situations as described in the Code. A tie connection between two transformer secondaries is shown in figure 28-3. Note that the tie conductor circuit has overcurrent protection at each end and that there are no load taps in the tie connections.

However, when load taps are made in the tie circuit between transformers, the minimum size of conductor required is regulated by the Code. In this case, the current-carrying capacity shall not be less than a stipulated percentage of the rated secondary current of the largest capacity transformer connected to the secondary tie.

A tie connection (with load taps) between two transformer secondaries is shown in figure 28-4. Since there are load taps present, the size of the tie conductors must be increased.

The overload protective devices used for the tie connection with load taps must be approved by the Underwriters' Laboratories, Inc. (UL). The following, when approved, are acceptable for protection:

- limiting devices consisting of fusible-line cable connectors (limiter);
- automatic circuit breakers.

If the voltage exceeds a value specified in the Code, the tie conductors must have a switch at each end of the tie circuit. When these switches are open, the limiters and tie conductors are deenergized. These switches shall be not less than the current rating

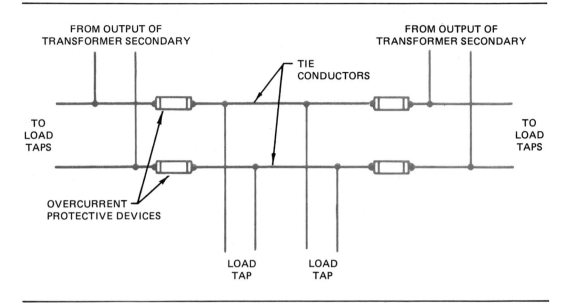

Fig. 28-4 Tie connections between transformers with load taps

of the tie conductors. Further, these switches shall be capable of opening their rated current.

The Code gives further information on overcurrent protection where secondary tie connections are used. It is acceptable practice to provide overcurrent protection in the secondary connections of each transformer. The setting of this overcurrent device is regulated by the Code.

An automatic circuit breaker must be installed in the secondary connection of each transformer. This circuit breaker must have a reverse-current relay set to open the circuit at not more than the rated secondary current of the transformer.

PARALLEL OPERATION OF TRANSFORMERS

Transformers may be operated in parallel and protected as a unit if their electrical characteristics are similar. These electrical characteristics include the voltage ratio and the percentage impedance. When transformers have similar electrical characteristics, they will divide the load in proportion to their kVA rating.

GUARDING OF TRANSFORMERS

Appropriate provisions must be made to minimize the possibility of damage to transformers from external causes. This is particularly important if the transformers are located where they are exposed to mechanical injury.

Dry-type transformers must be provided with a noncombustible moisture-resistant case or enclosure which will provide reasonable protection against the accidental insertion of foreign objects. They also should have weatherproof enclosures when installed outdoors. The transformer installation must conform with the Code provisions for the guarding of live parts.

The operating voltage of exposed live parts of transformers must be marked by warning signs or visible markings. These markings or signs are to be mounted in unobstructed positions on the equipment and structure.

GROUNDING

The National Electrical Code requires that the metal cases and tanks of transformers be grounded. Further, all noncurrent-carrying metal parts of transformer installations and structures, including fences, are also to be grounded. This grounding must be done in the manner prescribed by the Code to minimize any voltage hazard that may be caused by insulation failures or static conditions.

TRANSFORMER NAMEPLATE DATA

According to the Code, each transformer shall be provided with a nameplate and the nameplate must include the following information:

a. manufacturer's name;
b. rated kVA capacity;
c. frequency in hertz;
d. primary and secondary voltages;
e. amount of insulating liquid and type used; (This information is required only when the transformer rating exceeds 25 kVA.)
f. if Class B insulation is used in dry-type transformers rated at more than 100 kVA, it should be indicated on the nameplate;
g. temperature rise for this insulation system;
h. impedance (25 kVA and greater).

DRY-TYPE TRANSFORMERS INSTALLED INDOORS

Dry-type transformers are used extensively for indoor installations. These transformers are insulated and cooled by air. They are not encased in the steel tanks required for oil-filled transformers. For protection, dry-type transformers are enclosed in sheet metal cases with openings to allow air to circulate.

The Code specifies that dry-type transformers of a 112 1/2-kVA rating or less must have a separation of 12 inches from any combustible material. However, there are Code conditions and exceptions.

Some transformers of more than a specific rating must be installed in a transformer room with fire-resistant construction or must be installed in a transformer vault. Transformers with Class B insulation (80°C temperature rise) or Class H insulation (150°C temperature rise) need not be installed in a transformer vault provided they are separated from combustible material by the horizontal and vertical dimensions specified in the Code, or are separated from combustible material by a fire-resistant barrier. Any dry-type transformer rated at more than a high voltage specified by the Code must be installed in a transformer vault.

Fig. 28-5 Raising an oil-filled transformer to the roof of an office building *(Courtesy of the Los Angeles Department of Water and Power)*

Fig. 28-6 Oil-insulated transformer installed outdoors (A unit substation consisting of two 4,500-kVA, 11,500 to 2,300 to 460-V, three-phase, 60-Hz transformers) *(Courtesy of General Electric Company)*

ASKAREL-INSULATED TRANSFORMERS INSTALLED INDOORS

The windings of some transformers are cooled and insulated by a synthetic, non-flammable liquid called *askarel*. Askarel, when decomposed by an electric arc, produces only nonexplosive gases.

The Code specifies that some askarel-insulated transformers must be furnished with a pressure-relief vent. If this type of transformer is installed in a poorly ventilated area, it must be furnished with some method of absorbing gases that may be generated by arcing inside the case.

Any askarel-insulated transformer rated above a Code specified voltage must be installed in a vault.

OIL-INSULATED TRANSFORMERS INSTALLED INDOORS

Many transformers are cooled and insulated with a special insulating oil. The fire hazard potential due to oil-insulated transformers is greater than that of askarel-insulated transformers; therefore, the Code requirements are more exacting for oil-insulated transformers.

OIL-INSULATED TRANSFORMERS INSTALLED OUTDOORS

The Code requires that combustible buildings, door and window openings, and fire escapes must be safeguarded from fires originating in oil-insulated transformers. Such protection may be provided by effective space separation or by erecting a fire-resistant barrier between the transformer bank and the areas requiring protection.

In addition, the Code requires that some means be installed to dispose of the transformer oil from a ruptured transformer tank. Such a precaution applies to a transformer installation adjacent to a building where an oil explosion can result in a fire hazard without this preventive measure.

PROVISIONS FOR TRANSFORMER VAULTS

The Code regulations cover all essential details for vaults used for transformer installations, including the arrangement, construction, and ventilation of the vaults.

AUTOTRANSFORMERS

Code specifications are given for the use of autotransformers for lighting circuits. Recall that an autotransformer does not have separate primary and secondary windings. It consists of only one winding on an iron core. Part of the single winding of the autotransformer is common to both the primary and secondary circuits.

The Code limits the use of an autotransformer to feeding branch circuits because of the interconnection of the primary and secondary windings. The autotransformer may be used only where the identified ground wire of the load circuit is connected solidly to the ground wire of the source (NEC *Section 210-9*).

Figure 29-8 illustrates an autotransformer connected to a lighting load. Note that the ground wire is carried through the entire system.

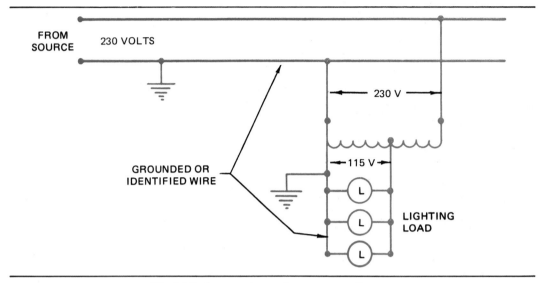

Fig. 28-7 Approved use of an autotransformer

An alternate use of an autotransformer for lighting circuits is shown in figure 28-8. This circuit also follows Code regulations as the identified ground wire is carried through the entire system.

Figure 28-9 illustrates an application for an autotransformer which is NOT approved by the Code. Here, the single-phase, 230-volt input to the autotransformer is obtained from a three-phase, 230-volt source. The use of a mid-tap on the autotransformer makes available a single-phase, three-wire system for a lighting load. However, if the ground wire is not solidly connected through the entire system, this circuit will

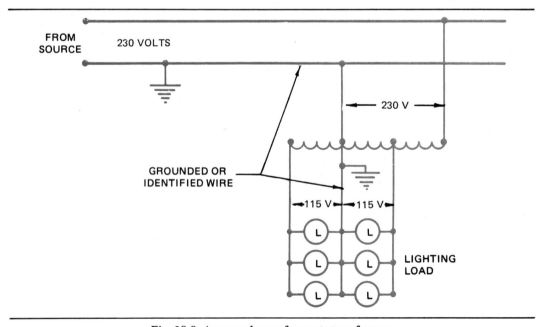

Fig. 28-8 Approved use of an autotransformer

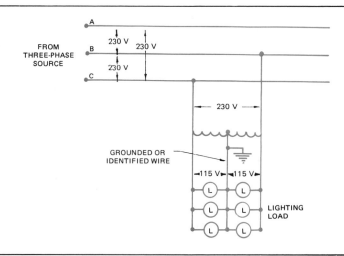

Fig. 28-9 Incorrect use of an autotransformer

Fig. 28-10 Voltage regulator transformer
(Courtesy of Westinghouse Electric and Manufacturing Company)

Fig. 28-11 Dry-type buck-boost transformer
(Courtesy of Hevi-Duty Electric, A Unit of General Signal)

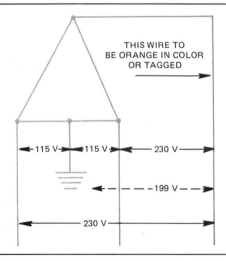

Fig. 28-12 Four-wire delta secondary identification

not meet Code regulations. The type of installation shown in figure 28-9 is unsafe, particularly if an accidental ground condition develops on the three-phase system.

"Buck" and "Boost" Transformers

Autotransformers are also used where only a small voltage increase (boost) or decrease (buck) is required, (figure 28-11). An example is to operate a 230-volt appliance from a 208-volt feeder line. This is accomplished by the use of an autotransformer which increases the 208-volt feeder line voltage to the 230 volts required to operate the appliance. Voltage drops (losses) in long or heavily loaded distribution systems may be increased in this manner as an energy conservation measure. Many regular (isolating) transformer corrections are used as autotransformers to decrease or increase a voltage.

IDENTIFIED FEEDER

On a four-wire, delta-connected secondary feeder conductor where the midpoint of one phase is grounded to supply lighting and similar loads, the phase conductor with the highest voltage to ground must be identified by an outer finish that is orange in color, or by tagging, (figure 28-12). This identification is to be placed at any point where a connection is made if the neutral conductor is also present, such as in a distribution panel, junction box, or pull box.

ACHIEVEMENT REVIEW

1. The rated primary current of a transformer is 4 amperes at 480 volts. How is the permissible maximum current setting determined for the overload devices used on the primary side of the transformer when secondary overcurrent protection is omitted?

   _____

   _____

   _____

   _____

2. The Code permits overcurrent protection in the secondary in place of the primary protection if two requirements for transformers 600 volts or less are observed. What are these requirements?

   a. _____

   _____

   b. _____

   _____

3. What items of data should appear on a transformer nameplate to comply with Code requirements?

   _____

   _____

4. What is a secondary tie circuit? _____

_____

_____

5. What happens to the secondary tie conductor size if loads are connected to the tie
conductors? _____

_____

6. What overload devices are acceptable to protect a secondary tie connection with
load taps? _____

_____

7. What are the electrical characteristics that must be similar if transformers are to
be operated in parallel? _____

_____

8. What are the Code requirements for the grounding of transformer installations?

_____

_____

9. How is the load divided when transformers are operated in parallel and protected
as one unit? _____

_____

_____

10. What precaution must be observed in using autotransformers to supply lighting
circuits which are grounded? _____

_____

_____

11. Give an example of how an autotransformer is used in a circuit to increase the
voltage. _____

_____

_____

12. Connect the following isolating transformer as an autotransformer to "boost" the
voltage from source to load.

208-V
SOURCE

H_2

208 V

H_1

X_1

24 V

X_2

230-V
LOAD

(232 V)

SUMMARY REVIEW
OF UNITS 21-28

OBJECTIVE

- To give the student an opportunity to evaluate the knowledge and understanding acquired in the study of the previous eight units.

1. What is a step-down transformer? _____

_____

_____

_____

_____

2. What is a step-up transformer? _____

_____

_____

_____

_____

3. List the items that should be marked on the nameplate of a standard power or distribution transformer. _____

_____

_____

_____

_____

4. Draw the schematic diagram of an additive polarity transformer.

5. Draw the schematic diagram of a subtractive polarity transformer.

For questions 6 through 17, select the correct answer for each of the statements and place the corresponding letter in the space provided.

6. The primary and secondary windings of an operating transformer are tied together _____
   a. electrically.
   b. magnetically.
   c. through switching gear.
   d. not at all.

7. The H leads of a transformer are connected to the _____
   a. high-voltage side.
   b. low-voltage side.
   c. secondary side.
   d. primary side.

8. The primary winding of a transformer is the _____
   a. high-voltage side.
   b. low-voltage side.
   c. input winding.
   d. output winding.

9. The single-phase, three-wire system is _____
   a. 115/230 volts.
   b. 120/208 volts.
   c. 230 volts.
   d. 120/277 volts.

10. An open-delta connection _____
    a. is the same as a closed delta.
    b. is an incomplete connection.
    c. requires three single-phase transformers.
    d. requires two single-phase transformers.

11. The line voltage of a three-phase delta system is the same as _____
    a. the line current.
    b. a single transformer voltage.
    c. a single transformer current.
    d. 1.73 X phase voltage.

12. The neutral of a three-phase, four-wire system is _____
    a. grounded.
    b. ungrounded.
    c. live.
    d. bonded.

13. A V connection is the same as the
    a. wye connection.   c. open-delta connection.
    b. delta connection.   d. open-wye connection.
    _____

14. Three 100-kVA transformers are connected in delta-delta. What is the total kVA capacity.
    a. 58 percent of the three ratings
    b. 58 percent of two ratings
    c. 100 kVA
    d. 300 kVA
    _____

15. Five amperes is the standard rating of a(n)
    a. instrument.
    b. secondary of a current transformer.
    c. secondary of a potential transformer.
    d. voltmeter movement.
    _____

16. It is dangerous to open the operating secondary of a
    a. closed-delta transformer circuit.
    b. open-delta transformer circuit.
    c. potential transformer.
    d. current transformer.
    _____

17. A transformer with part of the primary serving as a secondary is a(n)
    a. current transformer.   c. potential transformer.
    b. autotransformer.   d. open-delta transformer.
    _____

18. What are three standard types of cores used in transformers?

    a. _____

    b. _____

    c. _____

19. Name three common methods used to cool transformers.

    a. _____

    b. _____

    c. _____

20. A transformer has 1,200 turns in its primary winding and 120 turns in its secondary winding. The primary winding is rated at 2,400 volts. What is the voltage rating of the secondary winding?
    _____

21. State a simple rule to follow in connecting a transformer bank in closed delta.

_____

_____

_____

22. What is one practical application of single-phase transformers connected in a delta-delta configuration?

_____

_____

23. What is one practical application of an open-delta transformer bank?

_____

_____

24. The three-phase, four-wire secondary output of a wye-connected transformer bank can be used for two types of load. They are:

    a. _____

    b. _____

25. State a simple rule that may be used in connecting single-phase transformers in wye.

_____

_____

26. What are two practical applications for a three-phase, delta-wye transformer bank?

    a. _____

    b. _____

27. What is one practical application for a three-phase, wye-delta transformer bank?

_____

_____

28. What is one advantage of using a three-phase transformer in place of three single-phase transformers? _____

_____

_____

29. What is one disadvantage of using a three-phase transformer in place of three single-phase transformers? _____

_____

_____

_____

30. Insert the word or phrase to complete each of the following statements.

    a. A wye-delta transformer bank has _____ -connected primary windings and _____ -connected secondary windings.

    b. A delta-wye transformer bank has _____ -connected primary windings and _____ -connected secondary windings.

    c. A delta-wye transformer bank is used to _____ three-phase voltages to _____ values.

    d. A wye-delta transformer bank is used to _____ extremely high three-phase voltages.

    e. A three-phase transformer takes _____ space than a transformer bank of the same kVA capacity consisting of three single-phase transformers.

31. List the five common three-phase connections used to connect transformer banks consisting of either two or three single-phase transformers.

    a. _____

    b. _____

    c. _____

    d. _____

    e. _____

32. What are the two distinct types of instrument transformers?

    a. _____

    b. _____

33. Why must the secondary circuit of a current transformer be closed when there is current in the primary circuit?

    _____

    _____

    _____

34. The transformers shown in the following diagrams have a 4 to 1 step-down ratio. Determine the secondary voltage of each transformer. In addition, determine the value that each voltmeter will indicate.

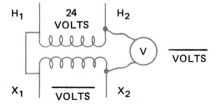
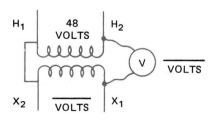

35. Determine the unknown values. What is the polarity of the transformer?

(Additive; Subtractive) _____

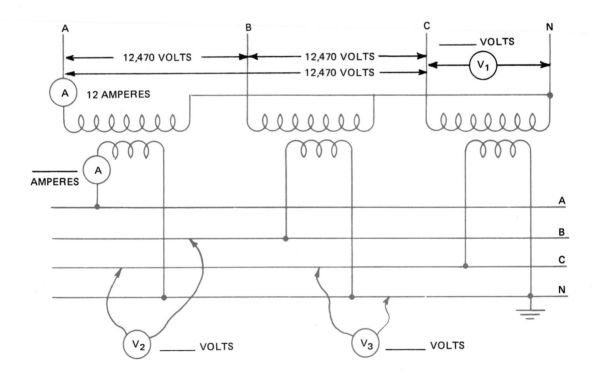

H₂

480 TURNS

240 VOLTS

_____ AMPERES

H₁

X₁

32 TURNS

_____ VOLTS

120 AMPERES

X₂

36. Two 75-kVA transformers are connected in open delta. Determine the total kVA capacity of the transformers. _____

37. A transformer is marked 37.5 kVA. Its primary is rated at 480 volts and its secondary is rated at 120 volts. Calculate the primary and secondary current rarings.

(Primary) _____

(Secondary) _____

38. Calculate the values that will be indicated by the ammeter and the three voltmeters shown in the following diagram. The transformer ratio is 26 to 1. Insert the answers in the spaces provided on the diagram.

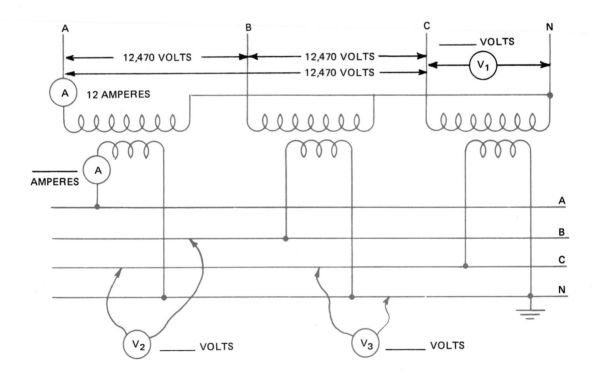

39. Explain the purpose of a buck or boost transformer. _____

_____

_____

_____

_____

40. On a four-wire delta system, what is an identified secondary feeder, and how is it identified? _____

_____

_____

_____

GLOSSARY

ALTERNATING CURRENT. A current which alternates regularly in direction. Refers to a periodic current with successive half waves of the same shape and area.

ARMATURE. A cylindrical, laminated iron structure mounted on a drive shaft. It contains the armature winding.

ARMATURE WINDING. Wiring embedded in slots on the surface of the armature. Voltage is induced in this winding on a generator.

AUTOTRANSFORMER. A transformer in which a part of the winding is common to both the primary and secondary circuits.

BLOWOUT COIL. Electromagnetic coil used in contactors and motor starters to deflect an arc when a circuit is interrupted.

BRANCH CIRCUIT. That portion of a wiring system that extends beyond the final overcurrent device protecting the circuit.

BRUSH POLARITY. Used to distinguish between the electrical polarity of the brushes and the magnetic polarity of the field poles.

BUCK OR BOOST TRANSFORMERS. Transformers used to boost (increase) a voltage or to buck (lower) it. These are small amounts of change.

BUSWAY. A system of enclosed power transmission that is current and voltage rated.

CIRCUIT BREAKER. A device designed to open and close a circuit by non-automatic means and to open the circuit automatically on a predetermined overcurrent without injury to itself when properly applied within its rating.

COMMUTATING POLES. Interpoles, energized by windings placed in series with the load circuit of a DC motor or generator.

COMMUTATOR. Consists of a series of copper segments which are insulated from one another and the mounting shaft; used on dc motors and generators.

COMPENSATOR TRANSFORMER. A tapped autotransformer which is used for starting induction motors.

COMPOUND-WOUND GENERATOR. A dc generator with a shunt and series, double field winding.

CONDUIT PLAN. A diagram of all external wiring between isolated panels and electrical equipment.

CONSTANT-CURRENT TRANSFORMERS. Used for series street lighting where the current must be held constant with a varying voltage.

CONTACTOR. An electromagnetic device that repeatedly establishes or interrupts an electric power circuit.

CORE-TYPE TRANSFORMER. The primary is wound on one leg of the transformer iron and the secondary is wound on the other leg.

COUNTER EMF. An induced voltage developed in a dc motor while rotating. The direction of the induced voltage is opposite to that of the applied voltage.

CUMULATIVE COMPOUND-WOUND GENERATOR OR MOTOR. A series winding is connected to aid the shunt winding.

CURRENT. The rate of flow of electrons which is measured in amperes.

CURRENT FLOW. The flow of electrons.

DELTA CONNECTION. A circuit formed by connecting three electrical devices in series to form a closed loop. Used in three-phase connections.

DIFFERENTIAL COMPOUND-WOUND GENERATOR. A series winding is connected to oppose the shunt winding.

DIODE. A two-element device that permits current to flow through it in only one direction.

DIRECT CURRENT (dc). Current that does not reverse its direction of flow. It is a continuous nonvarying current in one direction.

DISCONNECTING SWITCH. A switch which is intended to open a circuit only after the load has been thrown off by some other means; not intended to be opened under load.

DISTRIBUTION TRANSFORMER. Usually oil filled and mounted on poles, in vaults, or in manholes.

DOUBLE-WOUND TRANSFORMER. Has a primary and a secondary winding. These two windings are independently isolated and insulated from each other.

DYNAMIC BRAKING. Using a dc motor as a generator, taking it off the supply line and applying an energy dissipating resistor to the armature. Dynamic braking for an ac motor is accomplished by disconnecting the motor from the line and connecting dc power to the stator windings.

EDDY CURRENT. Current induced into the core of a magnetic device. Causes part of the iron core losses, in the form of heat.

EFFICIENCY. The efficiency of all machinery is the ratio of the output to the input.
$$\frac{\text{output}}{\text{input}} = \text{efficiency}$$

ELECTRIC CONTROLLER. A device, or group of devices, which governs, in some predetermined manner, the electric power delivered to the apparatus to which it is connected.

ELEMENTARY DIAGRAM (Ladder Diagram, Schematic Diagram, Line Diagram). Represents the electrical control circuit in the simplest manner. All control devices and connections are shown as symbols located between vertical lines that represent the source of control power.

FEEDER. The circuit conductor between the service equipment or the switchboard of an isolated plant and the branch circuit over-current device.

FLUX. Magnetic field; lines of force around a magnet.

FUSE. An overcurrent protective device with a circuit opening fusible part that is heated and severed by the passage of overcurrent through it.

GENERATOR. Machine that changes mechanical energy into electrical energy. It furnishes electrical energy only when driven at a definite speed by some form of prime mover.

GROUNDED. Connected to earth or to some conducting body that serves in place of earth.

HERTZ. The measurement of the number of cycles of an alternating current or voltage completed in one second.

HYSTERESIS. Part of iron core losses.

IDENTIFIED CONDUCTOR (Neutral). A grounded conductor in an electrical system, identified with the code color white.

INDUCED CURRENT. Current produced in a conductor by the cutting action of a magnetic field.

INDUCED VOLTAGE. Voltage created in a conductor when the conductor interacts with a magnetic field.

INDUCTION. Induced voltage is always in such a direction as to oppose the force producing it.

INSTRUMENT TRANSFORMERS. Used for metering and control of electrical energy, such as potential and current transformers.

INSULATOR. Material with a very high resistance which is used to electrically isolate two conductive surfaces.

ISOLATING TRANSFORMER. A transformer in which the secondary winding is electrically isolated from the primary winding.

LENZ'S LAW. A voltage is induced in a coil whenever the coil circuit is opened or closed.

MOTORIZING. A generator armature rotates as a motor.

NEC. National Electrical Code.

NORMALLY OPEN AND NORMALLY CLOSED. When applied to a magnetically operated switching device, such as a contactor or relay, or to the contacts of these devices, these terms signify the position taken when the operating magnet is de-energized, and with no external forces applied. The terms apply only to nonlatching types of devices.

OPEN DELTA. Two transformers connected in a "V" supplying a three-phase system.

OVERLOAD PROTECTION (Running Protection). Overload protection is the result of a device that operates on excessive current, but not necessarily on a short circuit, to cause the interruption of current flow to the device governed.

PARALLEL CIRCUIT. A circuit that has more than one path for current flow.

PERMEABILITY. The ease with which a material will conduct magnetic lines of force.

POLARITY. Characteristic (negative or positive) of a charge. The characteristic of a device that exhibits opposite quantities, such as positive and negative, within itself.

POLE. The north or south magnetic end of a magnet; a terminal of a switch; one set of contacts for one circuit of main power.

POLYPHASE. An electrical system with the proper combination of two or more single-phase systems.

POLYPHASE ALTERNATOR. A polyphase synchronous alternating current generator, as distinguished from a single-phase alternator.

POWER FACTOR. The ratio of true power to apparent power. A power factor of 100% is the best electrical system.

PUSH BUTTON. A master switch; manually operated plunger or button for an electrical actuating device; assembled into push-button stations.

RACEWAY. A channel, or conduit, designed expressly for holding wires, cables, or busbars.

RATING. The rating of a switch or circuit breaker includes (1) the maximum current and voltage of the circuit on which it is intended to operate, (2) the normal frequency of the current; and (3) the interrupting tolerance of the device.

RECTIFIER. A device that converts alternating current (ac) into direct current (dc).

REGULATION. Voltage at the terminals of a generator or transformer, for different values of the load current; usually expressed as a percentage.

RELAY. Used in control circuits; operated by a change in one electrical circuit to control a device in the same circuit or another circuit.

REMOTE CONTROL. Controls the function initiation or change of an electrical device from some remote place or location.

RESIDUAL FLUX. A small amount of magnetic field.

RHEOSTAT. A resistor that can be adjusted to vary its resistance without opening the circuit in which it may be connected.

SEMICONDUCTOR. Materials which are neither good conductors not good insulators. Certain combinations of these materials allow current to flow in one direction but not in the opposite direction.

SEPARATELY-EXCITED FIELD. The electrical power required by the field circuit of a dc generator may be supplied from a separate or outside dc supply.

SERIES FIELD. In a dc motor, has comparatively few turns of wire of a size that will permit it to carry the full load current of the motor.

SERIES WINDING. Generator winding connected in series with the armature and load; carries full load.

SHELL-TYPE TRANSFORMER (Double Window). The primary and secondary coils are wound on the center iron core leg.

SHIELDED-WINDING TRANSFORMER. Designed with a metallic shield between the primary and secondary windings; provides a safety factor by grounding.

SHORT AND GROUND. A flexible cable with clamps on both ends. It is used to ground and short high lines to prevent electrical shock to workmen.

SHUNT. To connect in parallel; to divert or be diverted by a shunt.

SHUNT GENERATOR. Dc generator with its field connected in parallel with the armature and load.

SILICON-CONTROLLED RECTIFIER (SCR). A four-layer semiconductor device that is a rectifier. It must be triggered by a pulse applied to the gate before it will conduct electricity.

SINGLE-PHASE. A term characterizing a circuit energized by a single alternating emf. Such a circuit is usually supplied through two wires.

SOLID STATE. As used in electrical-electronic circuits, refers to the use of solid materials as opposed to gases, as in an electron tube. It usually refers to equipment using semiconductors.

SPEED CONTROL. Refers to changes in motor speed produced intentionally by the use of auxiliary control, such as a field rheostat or automatic equipment.

SPEED REGULATION. Refers to the changes in speed produced by changes within the motor due to a load applied to the shaft.

STEP-DOWN TRANSFORMER. With reference to the primary winding the secondary voltage is lower.

STEP-UP TRANSFORMER. The secondary voltage is higher than the primary voltage.

THREE PHASE. A term applied to three alternating currents or voltages of the same frequency, type of wave, and amplitude. The currents and/or voltages are one-third of a cycle (120 electrical time degrees) apart.

THREE-PHASE SYSTEM. Electrical energy originates from an alternator which has three main windings placed 120 degrees apart. Three wires are used to transmit the energy.

THYRISTOR. An electronic component that has only two states of operation — on or off.

TORQUE. The rotating force of a motor shaft produced by the interaction of the magnetic fields of the armature and the field poles.

TRANSFORMER. An electromagnetic device that converts voltages for use in power transmission and operation of control devices.

TRANSFORMER BANK. When two or three transformers are used to step down or step up voltage on a three-phase system.

TRANSFORMER PRIMARY TAPS. Alternative terminals which can be connected to more closely match the supply, primary voltage.

TRANSFORMER PRIMARY WINDING. The coil that receives the energy.

TRANSFORMER SECONDARY WINDING. The coil that discharges the energy at a transformed or changed voltage, up or down.

UNDERCOMPOUNDING. A small number of series turns on a compound dc generator that produces a reduced voltage at full load.

VOLTAGE CONTROL. Intentional changes in the terminal voltage made by manual or automatic regulating equipment, such as a field rheostat.

WELDING TRANSFORMERS. Provide very low voltages and high current to arc welding electrodes.

WIRING DIAGRAM. Locates the wiring on a control panel in relationship to the actual location of the equipment and terminals; specific lines and symbols represent components and wiring.

WYE CONNECTION (Star). A connection of three components made in such a manner that one end of each component is connected. This connection generally connects devices to a three-phase power system.

INDEX